OBSERVING VISUAL DOUBLE STARS

OBSERVING VISUAL DOUBLE STARS

Paul Couteau

translated by Alan H. Batten

The MIT Press
Cambridge, Massachusetts
London, England

This translation copyright © 1981 by the Massachusetts Institute of Technology
Originally published in France under the title *L'Observation des étoiles doubles visuelles,* copyright © 1978 by Flammarion, Paris

All rights reserved. No part of this book may be reproduced in any form or by any means, electronic or mechanical, including photocopying, recording, or by any information storage and retrieval system, without permission in writing from the publisher.

This book was set in VIP Helvetica by DEKR Corporation and printed and bound by Halliday Lithograph in the United States of America.

Library of Congress Cataloging in Publication Data

Couteau, Paul, 1923–
  Observing visual double stars.

  Bibliography: p.
  Includes index.
  1. Stars, Double—Observers' manuals. 2. Stars, Double—Catalogs. I. Title.
  QB821.C7813   523.8'41   81-6070
  ISBN 0-262-03077-2   AACR2

CONTENTS

Foreword by J.-C. Pecker	vii
Preface	xiii

1
HISTORICAL INTRODUCTION — 1

2
OPTICAL CONCEPTS USEFUL TO DOUBLE-STAR OBSERVERS — 28

3
MEASURING INSTRUMENTS — 46

4
SOME PRACTICAL ADVICE — 77

5
THE IDENTIFICATION OF STARS — 92

6
COMPUTATION OF ORBITS AND STELLAR MASSES — 110

7
A VOYAGE TO THE COUNTRY OF DOUBLE STARS — 174

8
CATALOGUE OF 744 DOUBLE STARS FOR INSTRUMENTS OF ALL SIZES — 205

Index — 251

FOREWORD

A foreword has one great and rare advantage: It allows the writer, even if he is a friend of the author, the recognized right to enjoy the freedom of writing a few pages without any constraint; the freedom of a hammer without an anvil, or of a sketch without a design—unless that be to commend warmly a work read with sympathy, and to put forward some complementary ideas not included in the book. Therefore, this foreword will also, and indeed above all, be an appeal.

Paul Couteau's work begins with a review of the history of the discovery and measurement of visual double stars, from Herschel, who was the first to recognize the existence of the orbital motion of one member of a pair about the other, up to the latest (all too few) contemporary observers. After explaining the elements of classical optics, of diffraction theory and resolving power, necessary to understanding the problem posed, the author presents the measuring instruments and gives outlines of some practical "recipes" which, alone, enable a method that is *a priori* abstract and impersonal to be set on the road toward real success and the accumulation of useful results. The problems of identification, in particular, are difficult. What errors have novice observers not made because of this one problem! They would fill books themselves!

Chapter 6, on the computation of orbits and of stellar masses, gives the work its deep significance. It is not sufficient to accumulate measurements over the years, since the periods reach decades and sometimes exceed a century. The combined resources of trigonometry, celestial mechanics, and even spectroscopy (radial velocities measured by the Doppler-Fizeau effect of the displacement of spectral lines) must be deployed to compute the orbits and all their characteristics (period, size of the axes, eccentricity, inclination of the orbital plane to the line of sight, position of the nodes, etc.); and, especially, Kepler's laws must be used in the determination of stellar masses. Of course, the only accurately determined stellar masses come from this kind of work, and they must form the basis of the necessary calibration of *all* stellar masses—and galactic masses—in the universe. The neighborhood of the sun is a region where all the laws are calibrated, where all the values of standards are established, where all the methods of exploring the more distant parts of the universe are tested. Obviously, these observations, which may appear long and tedious, remain an essential element in contemporary astronomy and astrophysics.

In the last chapter (and in the remarkable catalogue that forms, in a sense, the conclusion) the author takes us on a voyage to the country of the double stars, based on the results of observations of past years and centuries.

Now that I have introduced the work, I would like to recall the personal memories that have kept me in touch, for more than a quarter-century now, with its author. Paul Couteau will not mind, I hope, if I recall these few memories. My only purpose in doing so is to try to answer the question, "How does one become an observer of double stars?" Paul Couteau, having received a good university education as a physicist, was attracted to astrophysics by his experience in amateur astronomy, which had been his hobby from childhood, pursued under the skies (sunnier and clearer than is generally believed) of the Vendée. He joined the Institute d'Astrophysique of the CNRS a few years after I did, and entered the same laboratory—that of Evry Schatzman. At that time, he worked on a thesis topic that combined Schatzman's interests with a proposal (no longer tenable) in my own thesis for a modern application of Schuster's mechanism for the formation of emission lines: How can the absorption lines of the white dwarf AC $+70°$ 8247 be explained? The working hypothesis was that the question was not one of unidentified absorption lines, but of spaces between hydrogen emission lines—those Balmer lines that we were astonished, at that time, not to observe in absorption as we did for all he other white dwarfs. Paul Couteau showed that to obtain the line intensities needed from the Schuster effect, effective temperatures of several hundreds of thousands of degrees were needed at gravities assumed to be very high. To be sure, this solution foresaw the high temperatures found today for the nuclei of planetary nebulae, but it posed many other problems; moreover, later, thanks to further progress in spectroscopy, the lines of the star in question were clearly identified. Several years after Paul Couteau's defense of his thesis, his working hypothesis, like thousands of others in astrophysics, was consigned to the ranks of those that are of "historic" interest.

Then Paul Couteau left us. A vacant post had to be filled at the Nice Observatory. André Danjon, then president of the Observatories Council, said to Couteau very bluntly: "Do you want this post? You can have it—but, a word of warning. Conditions are hard at Nice; you will be almost alone, certainly isolated, and I expect you to use the small equatorial to observe double stars. There is need for someone to do that." Paul Couteau, not frightened by the move or the thought of entering a field of which he knew little (which rather tempted him), took the bull by the horns, and, without flourishes or unnecessary demands, he left for Nice with packed bags, alone—a man from the Vendée thrown into the Alpes-Maritimes, into the olive groves of Mont-Gros, and into the atmosphere of *dolce fa niente* that was progressively invading the still wooded hills of an observatory in decay.

To be sure, he made friends down there, both with his colleagues and others, and he raised a large family in circumstances that bordered on the tragicomic. At last, however, he got the small equatorial (since named the Charlois equatorial) working, perfected it, and maintained it with the means at hand, and with very little help. With this instrument he observed many known double stars, discovered new ones, determined their orbits (with an old manual calculating machine), and soon became known in the international circle of double-star observers. He was regarded as a peer by van den Bos, van Biesbroeck, and Jonckheere, to mention only those who have died; the others are recalled in the first chapter of the book. He became, while still quite young, president of the Double-Star Commission of the International Astronomical Union, and he remains one of the most active double-star observers in the world.

I met Paul Couteau again when I myself became director of the

Nice Observatory in 1962. I had the good fortune to obtain enough financial assistance to give this branch of research, and Couteau in particular, the chance to use the entirely reconstructed great refractor of the Nice Observatory, as well as the Charlois equatorial, which was completely renovated at the same time.

Since that period, as before, his unwearying activity as an observer has led Paul Couteau to this book, a summary of his work of thirty years. Double stars, to be sure, are not one of the most "with it" areas of astronomical science, in the eyes of the majority of astronomers. Too many young researchers, who are seldom asked to give themselves to it, turn away to areas in which theses are more quickly completed—areas in which they can reap the harvest themselves. Few astrophysicists are aware that this is the work that produces the few stellar masses that they use. It is quite true that they seldom use masses obtained by direct analysis of an orbit; usually the application of the mass-luminosity relation is the source of their data. They ought not to forget, however, that the relation itself, even if it is (approximately) justified by theoretical arguments, can only be calibrated in practice by means of direct measures of double stars! It is often said that decades of observation are needed to obtain a good determination of an orbit and of a stellar mass—decades, however, of fundamental, essential work.

I recall this exemplary history to try to convince our young researchers (at least those who feel more at ease in front of an instrument than in front of a computer, and also, no doubt, many amateur astronomers) to turn themselves anew to this important but forsaken area of research. A few great observers of double stars are still active, but their ranks are thinning out. Paul Couteau belongs to the youngest generation. It is not certain that he will have successors; yet this is one of the most necessary areas of observation, and constitutes (together with the determination of parallaxes, proper motions, etc.) "equatorial astrometry." How I wish that my appeal may be heard!

The best introduction for potential observers in this field is Paul Couteau's book, which presents, in its more general astrophysical setting, the current state of work on double stars. It is a good, simple, clear, direct book. It is a work on which lots of young people should meditate, in proportion to the extent that many of them have lost those qualities of zeal that for a century made many astronomers devote their lives to a great work, to the making of a catalogue, to the using of a technique. Work, someone will say, of a Benedictine. Work of an astronomer in the highest sense of the term, which, every evening that the sky is clear, brings the observer to the dome, in front of the same stars, to make the same measures again and again; tedious but exciting, lasting decades. Work that brings him into the dome opened to the infinity of the night, alone with his micrometer, his eyepieces, his tables, his sidereal clock, and his notebook that he lights with a small pocket flashlight to jot down his measures. Work that leads him into his dome opened to the infinity of the starry night, with his woolly hat, his clumsy mittens, his padded pants. All alone with his marveling astronomer's eye. . . .

Jean-Claude Pecker
Membre de l'Institut
Professor at the College de France

PREFACE

This book is for all those who enjoy astronomy and who are particularly interested in the observation of double stars. It is as much for the young beginning amateur who wants to see for himself whatever can be seen with a small refractor as for the young assistant at an observatory who wants to try out several fields before making his choice, or for the experienced astronomer who will find in it precise and up-to-date data. It is accessible to everyone. It is the work I would have liked to have read when I first turned to the study of this kind of star in 1951. Much has been written on these objects, but we need a basic book. For a beginner in this area to obtain a clear idea of the state of this kind of research has been difficult.

The observation of double stars is richly fascinating—both for the amateur astronomer, because of the sights these stars afford, and for the professional, because of the problems they pose. To awaken a sense of vocation in other people, one must get across one's own love of the subject, then teach the method of research and show how it leads to a synthesis that can be a new point of departure. That is the aim of this book. It does not altogether resemble other books; it does not claim to be complete. It explains, in simple terms, the fruit of my experience of a quarter of a century in observation and research.

I owe my inclination toward double stars to André Danjon, who was the director of the Paris Observatory. It was necessary to bring back to life the old but powerful equatorials of Nice, to some extent abandoned during the war. The giant 18-m-long refractor—the fourth largest in the world—which I knew thanks to the works of Abbé Moreaux, had for a long time earned my admiration. When I read what Camille Flammarion had written in 1882 about refractors of aperture 10 cm, that "they enable one to verify by sight almost all the discoveries of modern astronomy," what could be said, then, about the great refractor of Nice, seven times bigger! This instrument, with which I was already familiar, was waiting for me, and, thanks to it, I felt that I could realize a childhood dream: to see objects that no one had yet seen.

I had the good fortune to be initiated by Robert Jonckheere, the last pioneer from the beginning of the century who was still active, and to have as models the other two Pauls of double-star astronomy: Paul Baize and Paul Muller. My luck continued, since I had at my disposition one of the four French 38-cm refractors (replaced after a few years by one of 50 cm, whose objective we owe to the farsightedness of André Danjon). At last, in 1969, the complete restoration of the great refractor, thanks to Jean-Claude Pecker, achieved the realization of the dream.

The book is divided into eight chapters. Each one treats some particular subject on the assumption that the reader does not know the material in the earlier chapters. At the end of each chapter is a partial bibliography.

The observation of double stars depends on people, and the history of the discipline is important if current procedures are to be understood. The first chapter gives the history. Next are to be found some very simple optical concepts that are, nevertheless, easily lost sight of. A double star is not observed in the same way as a planet or a variable star. How can one see a double star with very close components? The physiology of the observer and the theory of images must come together to produce a favorable result. The whole problem of masses depends on this. Different kinds of measuring instruments are studied; at least the ones that are or have been used. The filar micrometer is stressed. It has been used for 90% of all measures, and its principle is two centuries old. Some description is given of modern interferometric and occultation techniques which open up revolutionary perspectives.

Next, I discuss observational practice: the preparation for the night, the precautions to take with the mounting, the cunning struggle against the degradation of images by the atmosphere, and things to do or not to do.

Identifying stars is the subject of a whole chapter. This is a disappearing technique, like visual observation itself. To be sure, astronomers are used to identifying objects on plates, where they are to be found in hundreds; but the problem of identifying a single star seen through an eyepiece is different. That involves the whole history of star catalogues, which I deal with schematically, at best.

The classical chapter on the computation of orbits and of masses is addressed particularly to students. I describe only three methods of computing orbits, one of which is expeditious and provides the orbital elements within a few minutes. There is a table of masses carefully brought up to date, with references. The chapter finishes with the calculation of dynamical parallaxes. I stress the fact that I have tried not for completeness (the chapter would be too long), but for homogeneity. The computations appeal to basic notions of celestial mechanics. The reader will find assembled and synthesized all that is needed for the computation of individual stellar masses, from the computation of orbits from the observations to that of parallactic motion and of the center of mass. For each of these, numerical examples show how to proceed.

These classical concepts are used for the more original sequel, in which a voyage to the country of double stars is described. There I explain the research I have undertaken at Nice, and its results, in an imagery comprehensible to anyone with an elementary knowledge of astronomy and mathematics. Even the merely curious reader will understand how research should be organized. Where are the double stars? How would they be seen by inhabitants of these systems? How far can one see with a refractor? I answer by describing what we have done at Nice, and what follows from Paul Muller's and my success in discovering more than 2,000 double stars in ten years. I sort the stars into giants and dwarfs, and select those that should be observed each year, those that are close to us and whose study will permit us to understand the structure of these objects better.

The work concludes with an indispensable supplement: a catalogue of double stars. It is intended for everyone—from those who have only a simple pair of binoculars to societies that have large instruments. It is arranged so that each night the observer can find something suitable. It is limited to fairly bright stars that are easy to find. I have thought especially of those who do not have an equatorial mounting, as I do not during the summer in the country. Here the reader can find a new interest in observing double stars, or, better still, find his vocation as an observer (of which astronomy has so much need).

It remains to thank Jean-Claude Pecker, Professor at the College de France, for writing the foreword. Thanks to him, the Nice Observatory made a new start in 1963, and when he left us six years later the two great refractors of 76 cm and 50 cm, witnesses of our cooperation, enabled Nice to become one of the leading centers of visual observation.

I would like to thank my distinguished colleague Dr. Alan Batten for translating this book into English. Thanks to him, some important errors have been corrected and some ideas expressed more clearly for readers who are not professional astronomers.

OBSERVING VISUAL DOUBLE STARS

1
HISTORICAL INTRODUCTION

DOUBLE STARS IN RELATION TO THE UNIVERSE

The principal constituents of the universe are stars and nebulae. Stars are hot self-luminous objects composed mainly of hydrogen; they result from the division and contraction of gaseous nebulae under the influences of the forces of gravitation and radiation. The pressures generated by these contractions raise the gas to high temperatures. The stellar phase begins when the nuclear-fusion reactions that produce the star's own radiation are triggered deep in its heart. These objects come in many sizes. The sun, an average star, has a diameter of more than a million kilometers; the earth-moon system could easily fit inside it. Some stars are billions of times as large in volume, others billions of times smaller. [Note: 1 billion = 10^9.] By contrast, the range of masses is much narrower: from 1/20 to 40 times the mass of the sun (which, itself, contains 300,000 times as much matter as the earth).

Very deep down inside the sun, hydrogen is transformed into helium at a rate of 560 million tons every second. In relation to the sun's reserves of hydrogen, which will last for billions of years, this is a small quantity whose loss is hardly noticeable; but the energy released by the transmutation cannot but be noticed. The power it furnishes amounts to about 5×10^{23} horsepower. This energy is transmitted, little by little, to the surface regions, where it maintains a temperature close to 5,700°K; it escapes into space at a rate of 10 horsepower from each square centimeter of the surface of the sun (which is 10,000 times that of the earth). Some stars, such as S Doradus, radiate 300,000 times as much as the sun; others, such as Proxima Centauri, 100,000 times less.

Historical Introduction

Stars are grouped in giant clusters called galaxies. These are flattened systems so extensive that light, traveling at 300,000 kilometers per second, takes about 100,000 years to cross from one end to the other. They are often spiral in shape, and for that reason are sometimes called spiral nebulae. At present we reckon that there are 100 billion stars in a galaxy. Our sun belongs to one of these clusters called the Galaxy. It is a long way from the center (about 28,000 light-years), but almost in the plane of the system. The center of the Galaxy is in the constellation of Sagittarius, easily visible from our part of the world in the southern sky on summer evenings. The Milky Way is nothing but the light of the multitude of stars that make up the Galaxy. The stars in a galaxy are not stationary, but revolve around the center with speeds that are greater the closer they are to it. Thus, the sun carries us in an eternal round about the galactic center, but the nearby stars also take part in this motion, which long remained indiscernible even though its speed is close to 200 kilometers per second.

In our neighborhood, the average distance between two stars is so great that it takes four years for light to traverse it. The stars, then, are very far apart. If we could reduce them to the size of tennis balls they would be about 3,500 kilometers from each other. Their relative movement amounts to about their own diameter in a day. Clearly, collisions between stars are improbable.

Observation shows that, very often, stars come in pairs; these pairs are called *double stars*. The two bodies attract each other strongly, but this centripetal force is balanced by the centrifugal force of revolution. As a result, the two stars are in orbital motion about their center of gravity, just as the moon moves about the earth or the planets about the sun. It is obvious that the form and speed of this orbital motion can tell us about the masses of heavenly bodies. This is why double stars are important. Our sun is a single star; the retinue of planets that surrounds it would be invisible to astronomers like us observing from nearby stars. By contrast, our neighbor Alpha Centauri, in many ways like the sun (same age, same size), is double. A satellite star two or three times smaller accompanies it at a distance that light takes 4–5 hours to cover. These systems have probably been formed at the same time as the stars

3 Historical Introduction

they contain, by the contraction of a primordial nebula and its turbulent fragmentation. Thus, the study of double stars is important for cosmogony.

Double-star systems show great diversity. Some pairs of stars are so close to each other that their surfaces are almost in contact. Tidal effects are so great that the component bodies are drawn out into ellipsoids, and surface material is exchanged between them or even expelled, little by little, into the surrounding space. Such systems have very short periods of revolution (a few hours). A great number are known; for example, W Ursae Majoris is made up of two almost identical stars that revolve about each other in eight hours. Their centers are separated by less than 2 million kilometers, and their surfaces almost touch. Systems of this kind are recognized by analyzing their light with a spectrograph, and by studying their mutual eclipses, which make them vary in brightness (as does the star Algol). The distance between components is too small for us to see two bodies separately. These systems are called *spectroscopic* or *photometric* double stars, according to whether they are studied by spectrographs, which separate the wavelengths of light, or by photometers, with which the mutual eclipses may be measured.

When the components of a double star are farther apart (some hundreds of times their radii), they can be distinguished in refracting or reflecting telescopes. In this case they are called *visual* double stars. The observer is a spectator, watching the progress of orbits that require years and often centuries for their completion. He can see for himself that Newton's law of gravitation is indeed universal. He can check at the same time that Kepler's laws, which are derived from Newton's and control the motions of the planets, can be applied to stellar systems also, for as far as we can see. This fortunate state of affairs, observed in all double stars, demonstrates the universality of the laws of physics and allows us to intercompare the masses of the stars.

Sometimes the two stars in a system are identical, as in the pair γ Virginis, while in other systems, such as that of Sirius or that of α Ursae Majoris, the two bodies are very dissimilar. It can even happen that the companion cannot be distinguished at all, and that its existence is revealed by the motion of the primary star. Such systems are called *astrometric* double stars. The study of pairs made up of dissimilar bodies, called primary and secondary, gives us information about the evolution of stars. Eventually stars exhaust their supply of nuclear energy, and as they age their structure is modified; but the rate of aging is not the same for stars of different masses. Dissimilar components in a system do not evolve at the same speed. Thus, the study of double stars allows astronomers and physicists to tackle directly the problems posed by those mysterious bodies, the stars.

Double stars, or binaries, have been classified into several categories, according to the methods used in observing them. Sometimes one system belongs to more than one category. This is the case for α Centauri, a visual double star but also an astrometric and spectroscopic one. We shall consider in this book only the visual double stars; studying the other kinds requires very different techniques that are seldom accessible for amateurs. The distance between components and the inclination of the orbital plane with respect to the observer determine the category to which a system belongs. The very close stars, almost in contact, have very large relative speeds, which, owing to the Doppler-Fizeau effect, give well-displaced spectral lines, and the probability of eclipses is high. On the other hand, if the stars are well enough separated to be resolved, the relative speed is small and the couple cannot be studied with the spectrograph.

Systems of all sizes can be observed in the sky, right up to couples so widely separated that even the very weak attraction of neighboring stars can dissociate the system. Naturally, not all double stars are observable, so many remain to be discovered. Finally, there are chance associations, due to perspective, that are called *optical couples*. Their number is small beside that of the real double stars.

THE FASCINATION OF OBSERVING DOUBLE STARS

Double stars are so varied and so widely distributed that any instrument, even a simple pair of opera glasses, will show some. At present the number of known couples is increasing steadily; it now exceeds 60,000. One can still find systems by searching the sky and inspect-

5 Historical Introduction

ing stars one by one with the high magnifications associated with the great refractors and reflectors. Each new system contributes to our knowledge of the behavior of stars.

When refracting telescopes were invented, at the beginning of the seventeenth century, no one knew either the distances or the sizes of the stars. The idea of a double star did not originate with Galileo, the first astronomer to have a telescope. The first binary discovered was Mizar, second star in the handle of the Big Dipper, whose companion, nearly equal in luminosity and 14 arc-seconds away, was observed by Riccioli in 1650—41 years after Galileo's first observations. Some years later, in 1656, Huygens noticed the Trapezium in Orion, a sextuple system. These chance discoveries did not make people think of double stars, any more than did the discovery of the eclipsing binary Algol, in 1699, by Montanri. The close pairs were attributed to perspective, while the true nature of Algol, a system of two dissimilar stars that eclipse each other every 68 hours, was not established until two centuries later. It was William Herschel who first thought of double stars as physical systems, and who began research on them a century after Galileo.

The observation of double stars has always been a lively interest of numerous amateurs and professionals. Even an experienced astronomer, after many years of observation, cannot remain indifferent to the sight of one of these objects in the field of his instrument. Even in the case of a difficult couple that can be resolved only by large instruments, the fascination is not only in the sight but in the motion, more or less rapid, of the components about their center of gravity, which allows us, after the elapse of some time, to determine the relative orbit of one star with respect to the other.

Double stars are so diverse and numerous that an observer has a wide choice of objects with even the smallest refractor. In a small instrument the images are always sharp and stable, which is rarely the case for a big one. Observation of binaries allows one to test the quality of an objective; certain objects, such as Antares in Scorpio or ϵ Bootis, are difficult pairs. To see them well with an objective of 7 centimeters is already quite a feat. With an objective only slightly larger, 10 cm for example, the choice of objects becomes much wider, and pairs such as ξ Bootis, whose two equally bright components are separated by 1.1 arc-seconds, are indicators of the quality of an instrument. Some pairs with a great difference in brightness between the components (such as Sirius and its 10,000-times-less-luminous companion) that were discovered with powerful instruments are observable with refractors of quite modest aperture. The companion of Sirius has been seen at Nice with a 10-cm refractor, although it is difficult to observe even with very large instruments.

If you treat double-star observation as a game for a while, you will quickly become an enthusiast for these stars, and you will graduate to more and more difficult and interesting pairs. This kind of observation is very subjective. Burnham in America and Jonckheere in France discovered, with the modest means of amateurs, double stars that are very difficult to see even with much bigger instruments. For this reason, these extraordinary observers were for some time dismissed as pure fantasists.

The spectacle offered by bright and diversely colored double stars was described by authors of the last century, such as Arago and Flammarion, in the flowery style suited to the readers of their popular astronomies. These astronomers, helped by energetic observers, popularized the great refractors and favored their development by wealthy donors, thus rendering a decisive service to the whole of astronomy. It is to double stars that we owe the great 91-cm refractor financed by Lick in California, that of 69 cm at Bloemfontein, financed by Lamont, and (to some extent at least) the 101-cm refractor at Yerkes Observatory and the 76-cm Bischoffsheim refractor at Nice.

I will not give here a complete history of double stars, since that has already been done by several authors, such as Paul Baize (1930), Robert Grant Aitken (1935), and Wulff D. Heintz (1971). I will emphasize the important achievements and the recent history that are the sources of our knowledge of stellar couples.

In this realm the past is of great importance, because the observations needed for knowledge of the orbits often go back very far, more than a century. It is essential to know which astronomers observed the stars for which we now calculate orbits, and what equipment they used. The old observations are precious; the passage of time increases their scientific value.

7 Historical Introduction

THE FORERUNNERS: HERSCHEL, FATHER AND SON

Sir William Herschel (1738–1822) was the first to undertake the recording of double stars, beginning in 1776. His purpose was to measure stellar parallaxes by comparing the position of the primary star with that of its fainter neighbor, supposed to be much farther away along the same line of sight. This ingenious method, conceived by Galileo, eliminates the errors due to precession, nutation, and refraction, as well as numerous instrumental errors.

William Herschel observed with reflecting telescopes that he made himself, one of aperture 50 cm and another of 1.30 m; the mirrors were of speculum metal. These instruments, the most powerful of his time, enabled him to carry out successfully his famous star-counting observations, a task that could not be considered by the national observatories, which were preoccupied with work in positional astronomy undertaken for the purposes of navigation and mapmaking. The mirrors, made with the greatest care, gave well-rounded images; by contrast, the mountings were altazimuth and had no drives. The observer followed the star with his eye, sometimes without any eyepiece, from one side of the field to the other, moving the instrument from time to time by pushing it with his arms. Herschel's coarse micrometers did not allow him to make precise measures. The theory of stellar images was developed by Sir George B. Airy in 1850; in Herschel's day it was thought that the apparent diameter of an image depended on the magnification and the brightness. Since most of William Herschel's measures of double stars were made by comparison with an estimate of the size of the stellar disk, they have only small scientific value; but the passage of two centuries has made them very precious.

William Herschel realized that the motions observed in stellar systems were caused not by parallax but by orbital motion. At the beginning of the nineteenth century he published a memoir in which he described the first orbits outside the solar system, in particular those of Castor and γ Virginis. Thus, the history of double stars began as a by-product of research on the parallaxes of stars.

Sir John Herschel (1792–1871), William's son, resumed the study of double stars in 1816, in collaboration with Sir James South—first in the northern hemisphere, then in the southern at the Cape of

Good Hope. Their harvest of more than 3,000 couples is of great historic value because it showed the importance of the population of double stars and stimulated further research.

THE FIRST MODERNS: STRUVE, FATHER AND SON

If the Herschels were the initiators, their true successor Wilhelm Struve (1793–1864) is the pioneer of the modern astronomy of double stars. This astronomer was the first to have available (in 1824, at Dorpat Observatory in Estonia) an equatorially mounted refractor driven by a clock and equipped with a filar micrometer. This refractor of aperture 24 cm, constructed by Johann von Fraunhofer, was the most powerful of its time. Sketches of the instrument show all the essential parts of modern astronomical refractors.

Wilhelm Struve discovered 3,134 stellar pairs with this instrument. It was necessary to complete these discoveries by precise measurement of the objects on the sky. The Herschels, engrossed in many projects, had left the heavy task of making catalogues of positions to the national observatories. Wilhelm Struve, at the head of a large and well-equipped establishment, set out to fill this gap. At the time, only the positions of the brightest stars had been measured precisely with the meridian circle. Each new double star discovered was, in general, a new object without preliminary identification. It was of the first importance to have a precise position for it on the sky. Struve published his results in three fundamental works written in Latin: *Catalogus novus stellarum duplicium* (1827), which gives the list of his double stars with an approximate position and a brief description, *Stellarum duplicium et multiplicium mensurae micrometricae* (1837), in which may be found details of the measures of the pairs, and *Positiones mediae* (1852), which gives precise positions measured with the meridian circle.

Wilhelm Struve's method of work differed from that of his predecessors. His instrument, furnished with a drive and slow motions, freed him from the severe constraint of following stars by hand, which would necessarily have impaired his visual acuity. It is not necessary to seek any further for the reason why Herschel could not detect the companion of Sirius around 1815, or many other pairs that should have been obvious with the apertures he had available.

Wilhelm Struve observed up to 400 objects an hour. This is very quick; he had only nine seconds on average to locate his star in the finder, center it, and examine it with an adequate magnification. His aim was to examine the greatest possible area of the sky. In three years he examined 20,000 objects, discovering one double star for every 38 single ones. In 1839, after these researches, Wilhelm Struve founded Pulkovo Observatory, near St. Petersburg, and installed there a 38-cm refractor. With this instrument, Otto Struve (1819–1905) continued his father's work by discovering over 500 pairs, the measures of which were published in 1843 and 1850, and by making numerous observations of the stars of the *Catalogus Novus*. This 38-cm refractor dethroned the 24-cm from the first rank, and was surpassed itself only 27 years later by the 47-cm refractor of Dearborn Observatory at Northwestern University in Evanston, Illinois. But these instruments of Dorpat and Pulkovo served as models for the giant refractors of Europe and America of the end of the nineteenth century. The work of the Struves has become of great importance through the passages of a century, since 23 percent of the stellar orbits now known are of systems discovered by them. After this early research at Pulkovo, the history of double stars followed that of the large astronomical refractors.

THE AGE OF THE GREAT PIONEERS

The age of the great American pioneers began in 1873 with the first discoveries by Sheldon W. Burnham (1838–1921). This amateur astronomer quickly acquired an international reputation by discovering his first pairs with his own 13-cm refractor, with which he eventually made 437 discoveries. He was entrusted with the refractor of Dearborn Observatory, and there he found 409 couples, in collaboration with G. W. Hough, who discovered 622. Burnham's gifts as an observer and his enthusiasm complemented those of Lowell and Schiaparelli, who believed they had discovered life on the planet Mars and described the works of Martian engineers. The enthusiasm was communicated to wealthy donors who, probably fed up with the show put on by the human race, financed the great refractors of Europe and America. Burnham visited Lick (where he observed 248 pairs with the large refractor), Yerkes (61 pairs), Washington (14 pairs with the 66-cm refractor) and Washburn near Chicago (87 pairs with the 39-cm refractor). He discovered 1,336 double stars in all. Burnham looked for his pairs somewhat at random; his habit of examining stars near those he had measured opened the search to more systematic examinations.

At the beginning of the twentieth century, Burnham gathered in one large catalogue all the double stars known at that time. This remarkable work, *A General Catalogue of Double Stars within 120° of the North Pole,* published in 1906, gives the measures of, the identifications of, and a multitude of notes on each of 13,665 stars, and includes bibliographic data. It is a model of its kind for the clarity of its exposition, the abundance of its information, and the compiler's care for the truth. He verified for himself—at the telescope, on the sky—everything that appeared to him doubtful or incomplete.

After 1901 Burnham no longer looked for new couples, but he did a lot more work with the great refractor at Yerkes; from 1907 to 1912 he made 9,500 observations of pairs of large separation, with a view to determining their proper motions.

By his researches, Burnham had shown that the survey of visual binaries was far from complete. With the commissioning of the giant American refractors and the improvement of star catalogues, systematic searches could begin. For such work there are two ways to proceed: either by making celestial charts with the aid of catalogues that are complete to a certain magnitude, or by going directly from the catalogue to the sky. Robert G. Aitken (1864–1951) and William J. Hussey (1862–1926) prepared charts down to magnitude 9.0 and 9.1, with the help of the catalogues of the *Astronomische Gesellschaft,* and reviewed the sky star by star. The observations were begun in the autumn of 1899, with these two observers sharing the two refractors (30 cm and 91 cm) of the Lick Observatory. The small instrument, easier to maneuver than the larger, enabled them to inspect a greater number of stars per hour. More than a third of the couples discovered at Lick were discovered with this instrument, but almost all were measured with the larger one. Unfortunately, neither Aitken nor Hussey indicated which celestial zones were surveyed with which instrument, apart from the polar cap above declination 60°, which was observed only with the small refractor. There are, then, sizeable areas of the sky for which we do not know whether or not they have been studied with an instrument of aperture greater than 30 cm. This explains in large measure the present harvest of new couples.

At the end of five years, Aitken and Hussey had discovered 2,000 couples. But Hussey was appointed director of the observatory at Ann Arbor, Michigan, and Aitken continued alone. The survey was finished around 1915, with a total yield of 4,000 binaries, most of them difficult and many with short periods. Hussey, busy with his new duties, did not give up his research on binaries. At La Plata Observatory in Argentina, which was then administered by the University of Michigan, he discovered 300 pairs in the southern hemisphere between 1912 and 1914 with the 43-cm refractor. Then he devoted himself to the installation of the 69-cm refractor of Bloemfontein Observatory, which is named Lamont-Hussey after the donor and the scientist. Hussey was preparing to take up his work of discovery with this instrument when he succumbed to a heart attack a few days before the inauguration (October 1926).

Europe did not remain inactive. The Reverend T. E. H. Espin (1858–1934) in England, at his own observatory of Tow Law, with reflecting telescopes of apertures 45 and 60 cm, and Robert Jonckheere (1889–1974), first near Lille (33-cm refractor) and then at Greenwich (71-cm refractor), also discovered close to 4,000 double stars by direct examination. But these couples discovered with the less powerful instruments are statistically less interesting than those discovered in California or at Chicago, because very few have shown any orbital motion since their discovery some time ago.

Nevertheless, around 1925 the number of observations was such that a new general catalogue was needed. It was begun by E. Doolittle and finished and published by R. G. Aitken (1932) under the title *New General Catalogue of Double Stars within 120° of the North Pole*. This work is in two volumes, contains 17,180 objects, and gives for each pair all the measures published since Burnham's catalogue. Despite its age, this compilation still serves as a reference for observers because it has not been followed by any other up-to-date general catalogue.

THE MODERN SURVEYS

Apart from the survey by Espin and Milburn in England, which was continued until 1930, large surveys ceased in the northern hemisphere after 1918. It was thought at that time that the recording of visual binaries was almost finished. Jonckheere did not hesitate to write in the introduction to his catalogue published in 1917: "The rapid increase of new double stars has this year come to a sharp conclusion, at least for stars as bright as B.D. 9^m0. All these stars have now been examined at the Lick Observatory, and it is not probable that many of the stars whose duplicity could not be observed from Mount Hamilton will be detected elsewhere." However, Jonckheere did not say that many parts of the sky had been surveyed with the 30-cm refractor, which left numerous couples to be discovered even with average instruments.

After 1920, progress in stellar classification allowed binaries to be sorted according to spectral type, absolute magnitude (giants or dwarfs), and distance from the sun. This led Gerard P. Kuiper, in 1934, to make a survey according to astrophysical criteria. He examined the stars close to the sun with the large refractor at Lick and added 117 couples, some of which (especially the red dwarfs) proved to be remarkable for the rapidity of their revolution and provided knowledge of the masses of several stars within a few years. Kuiper's work clearly showed that the sky had not yielded to observers all the double stars that were accessible, especially among the dwarfs, as Burnham had noted 30 years earlier. Jonckheere took up his survey again at Marseille (80-cm reflector) with the aim of discovering more dwarf pairs, and, between 1941 and 1945, recorded more than 2,000 faint pairs. But he did not have the luck of Kuiper, who had limited himself to nearby stars, since a faint star is not necessarily a dwarf. Jonckheere's lists contain principally distant giants of very slow motion. The angular separation of these stars is relatively large, as are those of the binaries discovered by Espin.

To find dwarf binaries, it is necessary to examine only stars recognized as dwarfs. This is what Charles E. Worley did in 1960, using the lists of dwarf stars discovered spectrophotometrically by A. N.

Vyssotsky. In this way he found about 30 pairs of red dwarfs, of which about 10 are notable for their rapid orbital motion.

These researches show that the earlier surveys were not exhaustive. The quality of telescopic images, affected by atmospheric conditions just as much as by the power of the instrument, limits the yield of a survey. Some binaries that are difficult to see, for example those with a large difference of brightness between the components, will not be seen under the mediocre conditions with which the astronomer must often be content. Moreover, in the course of its orbital revolution a system can appear single on several occasions and escape even a careful examination. The result is that there are always many very interesting unknown binaries.

This is the reason why the Nice Observatory undertook a survey 10 years ago. A preliminary sounding with the 38-cm refractor among stars of large proper motion in the Paris zone immediately allowed us to discover 145 couples out of 5,250 stars examined. These encouraging results prompted the author to make another general survey with the 50-cm refractor installed in 1967 and the 74-cm refractor restored in 1969. This survey began at declination $+17°$ and continued to the north. From $17°$ to $32°$ the stars of the AGK3 were surveyed; then, beyond $32°$, those of Argelander's catalogue (the BD). In 9 years, 65,000 stars were examined and the number of discoveries increased to 1,500 binaries. For his part, Paul Muller, with the 50-cm refractor, examined the polar cap previously surveyed by Aitken with the Lick 30-cm refractor. North of $+60°$, the examination of 24,000 stars led to the discovery of 550 binaries. Thus, a total of more than 2,000 pairs has been discovered at Nice from 1967 to 1976. On average, examination reveals one double star for 50 single ones. This proportion is remarkably constant, both in right ascension and in declination. In 1976, the survey was 65 percent complete between the North Pole and declination $17°$. These new double stars, discovered on nights of excellent quality, are in general objects that are very difficult to see, even with very large instruments. At small separations, the number of pairs discovered at Nice is greater than the number previously known.

At the Belgrade Observatory, a search for new pairs is in progress with the 65-cm refractor. About 100 binaries have been recorded since 1969.

RESEARCH IN THE SOUTHERN HEMISPHERE

The southern hemisphere was surveyed much later. After the first discoveries at the Cape of Good Hope by John Herschel at the beginning of the nineteenth century, some observations were made at the Sydney Observatory. However, the resumption of observations of double stars, with modern methods, had to wait until the end of the century.

R. T. A. Innes (1861–1933), with a 45-cm refractor at the Cape, discovered 450 binaries between 1896 and 1903, and in 1899 published a first catalogue of double stars south of the equator. In 1903 he was appointed director of the Johannesburg Observatory, but he had to wait for the 67-cm refractor to be put in service in 1925 to increase his total to 1,600 couples. This instrument, followed in 1928 by the 69-cm refractor of the Lamont-Hussey Observatory in Bloemfontein, marked the beginning of a very large harvest of binaries. W. H. van den Bos (1896–1974) and W. S. Finsen (1905–1979), at Johannesburg, using the direct-catalogue method, discovered 3,200 pairs between 1925 and 1935, while R. A. Rossiter and his colleagues at Bloemfontein, using the chart method, discovered nearly 8,000 between 1928 and 1946—a tremendous effort.

As of 1946, it could be said that, thanks to these energetic observers, the southern sky had been better surveyed than the northern. In equal areas of the sky, about twice as many close pairs are known in the southern hemisphere as in the northern. There has been no supplement to Innes's first catalogue. Pairs south of declination $-19°$ have been recorded since 1927 by Innes and his successors in the *Southern Catalogue Looseleaf Mimeograph.* This is a card index perpetually kept up to date and known by the initials SDS (Southern Double Stars). Systematic surveying in the southern hemisphere stopped in 1950.

DOUBLE STARS DISCOVERED BY PHOTOGRAPHY

A large number of double stars have been found, not in the sky, but on photographic negatives originally obtained for the *Carte du Ciel* (see chapter 5). In this way, 15,000 pairs have been "discovered" without the investigator leaving the library. Many of them, however, do not really exist but are merely spots or bubbles on a plate. The negatives of the *Carte du Ciel* were not made for the study of double stars; the very small angular separation of the components generally prevents the impression of distinct images. Pairs discovered in this way are widely separated and do not show any appreciable relative motion. The situation is different with plates obtained at the focus of instruments of long focal length, with special precautions such as are practiced at Sproul (near Philadelphia) for the purpose of measuring stellar parallaxes, and with those obtained at Flagstaff with the 150-cm astrometric reflector. The number of astrometric binaries studied in this way is small, but some of them (such as Ross 614, which contains extreme red dwarfs) are of great interest.

DOUBLE STARS WITH LARGE COMMON PROPER MOTION

It has been evident for some time that there are a large number of dwarf stars, among which some pairs are known (principally through Kuiper's research). Visual observation does not allow us to reach a great number, because their faintness prevents us from seeing those beyond a distance of 25 light-years. By photography, however, it is easy to detect extreme dwarfs up to 500 light-years away.

Large numbers of dwarf doubles can only be reached by photography, and then only the widest couples, recognizable by the common proper motion of the components. Willem J. Luyten has contributed most to the discovery of faint double stars, using the Schmidt telescope on Palomar. A thirtieth of the sky has been surveyed, giving a total of 2,000 couples brighter than magnitude 21.5, of separation greater than 2 arc-seconds, and of common proper motion greater than 0.2 arc-seconds per year. Among these 2,000 dwarfs, 120 of the companions are white dwarfs or degenerate stars and 14 of the primary components are white dwarfs. Luyten concludes that these binaries are perhaps the most common type, but that because of their slow orbital motions we will never know their orbits.

SOME GREAT OBSERVERS OF DOUBLE STARS

The pioneer work done by the surveyors of double stars would be useless without its complement, which is continued observation of the pairs recorded in order to determine their orbits. Some surveyors follow up only their own discoveries, but some great observers in double-star astronomy have devoted themselves to observing and measuring carefully the pairs discovered by others. We should cite some of the most active. It often happens that these observers are not professionals, but dedicated amateurs, like Burnham, whose care for accuracy and capacity for work are their dominant characteristics. In the nineteenth century two Europeans distinguished themselves in this way. They were the Reverend W. R. Dawes (nicknamed "eagle eye"), in England, and Baron Ercole Dembowski in Italy. Each of these men, with small refractors not surpassing 15 cm in aperture, measured all the double stars accessible to their instruments with an accuracy now seen to be very valuable a century later.

In America, Georges van Biesbroeck (1880–1974), of Belgian origin, observed the most difficult couples during three-quarters of a century, from 1901 to 1974. In 1915 he had placed at his disposal the largest refractor in the world (the 101-cm at Yerkes), from 1939 he used the 2-m reflector of the McDonald Observatory in Texas, and from 1964 he used the new 2.1-m telescope at Kitt Peak in Arizona. Numerous orbits of double stars are due to the observations of this astronomer, who made 35,000 measures of 10,000 couples. Charles E. Worley at Washington and W. D. Heintz at Sproul continue the American tradition.

France has always played a leading role in the astronomy of double stars. Camille Flammarion (1842–1925) published in 1878 a *Catalogue des étoiles doubles et multiples en mouvement relatif certain,* a work remarkable for quality of production, care for detail, and the total effort that it represents. France has always had observers, too. Perrotin, in the 1880s, at the 38-cm refractor of Nice, and Giacobini, after the first world war, at the 30-cm refractor of Paris, made major contributions. Paul Baize, a physician by profession but a dedicated amateur astronomer, observed double stars without interruption from 1924. From 1930 to 1972, at Paris, he had

17 Historical Introduction

Figure 1.1 The Bischoffsheim dome, which houses the large refractor at Nice Observatory. Architect: C. Garnier. Dome by G. Eiffel. Nice Observatory photograph.

18 Historical Introduction

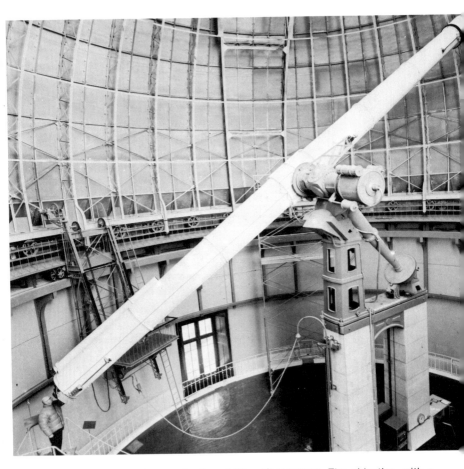

Figure 1.2 The large refractor of Nice Observatory. The objective, with a clear aperture of 74 cm, was reground in 1966 by J. Texereau. The focal length is 17.89 m and the resolving power is 0.16 arc-seconds. Nice Observatory photograph.

the use of refractors of 30 cm and then of 38 cm, with which he made nearly 25,000 measures. He is the only Frenchman to have measured binaries throughout the second world war. We owe to him, besides, more than 130 orbits and a catalogue of double red dwarfs. V. Duruy, a mining engineer and another dedicated amateur, has made thousands of observations with his own equipment at Nancy (27-cm refractor) and at Rouret in the Alpes-Maritimes (40-cm and 60-cm reflectors). The French have always been in the vanguard of progress in the improvement of micrometers used for the measurement of double stars. A. Danjon invented the half-wave-plate micrometer (1937), P. Muller a double-image micrometer (1937), and V. Duruy a comparison-star micrometer (1935).

I should also mention Rabe, in Germany, who made 30,000 measures from 1915 to 1956 with the 27-cm refractor at Munich. In Belgium, S. Arend and Jean Dommanget worked with a 45-cm refractor. In the southern hemisphere, J. Voûte, of Dutch origin, observed some thousands of couples from 1940 to 1944 at the Lembang Observatory in Java with 37-cm and 60-cm refractors.

Recently, some of the great observers have died: van Biesbroeck, van den Bos, and Jonckheere in 1974, all within a few months. Wide gaps are opening in the already sparse ranks of the observers, who can at present be counted on the fingers of one hand. The retirement of several others, such as Baize, and the death of Finsen, make the recruitment of observers urgent. It is to be hoped that double-star astronomy will find its devotees from among the 3.5 billion individuals inhabiting the globe, as it always has found them, and that the torch carried so well until now is not about to be extinguished.

CENTRALIZATION OF MEASURES

The results of the efforts in the southern hemisphere have never been assembled in a work like Aitken's 1932 catalogue. Moreover, that work has aged very quickly because of the increase in measures and in new pairs. The preparation of a modern edition of all the observations of double stars would prove onerous, and the new version would also be subject to rapid aging. It has seemed preferable to publish an index. Thus, all the known pairs, as well as all the measures of each, have been assembled on punched cards, each measure on one card. This enormous labor was completed at the Lick Observatory in 1963 by Hamilton M. Jeffers, W. H. van den Bos, and Frances M. Greeby. All the couples thus identified on cards have been brought together in one work: the *Index Catalogue of Visual Double Stars 1961.0.*

The *Index,* in two volumes, identifies 64,247 doubles. For each pair it gives the positions for the years 1900 and 2000, the dates of the first and the last measure, the number of measures made at various epochs (if the number exceeds 25, the symbol "99" indicates that the measures are numerous), the position angles and separations corresponding to the epochs of the first and last measures, the magnitudes of the components reduced to the scale of the *Henry Draper Catalogue,* the spectral types, the proper motion of the primary, the number in Argelander's catalogue (the BD) (or the Cordoba catalogue, for southern stars), the number in Aitken's catalogue, and finally a reference to explanatory notes. This work will be brought up to date about every 10 years. It serves now as a reference for the identification of all the couples published, without omission even of those whose existence is very doubtful.

The cards, numbering about 300,000, are now centralized in Washington; this is the *central card index.* Copies at Herstmonceux and Nice are brought up to date several times a year by the office of the central card index, for which C. E. Worley has assumed responsbility. The contents of the cards and of the *Index* have been put on magnetic tapes. One of the centers can send to any inquirer the measures he needs, so the periodic publication of large collections of measures is no longer necessary.

FEATURES OF THE CLASS OF KNOWN DOUBLE STARS

The 66,000 pairs harvested by our means of observation are not necessarily representative of the actual population of binaries among the stars. The following table gives the percentages of binaries in three luminosity intervals.

	Magnitude		
	<8	8–11	>11
Percentage of pairs	8%	65%	27%

Three-quarters of the known double stars are brighter than magnitude 11; they are, then, relatively bright. This gives about one pair for every 15 stars. The real ratio must be greater, because in the neighborhood of the sun it is close to 1/2. This shows that the majority of pairs escape detection because they are so far away.

An effect of observational selection becomes evident when it is noted that a third of the known couples are formed of nearly equal stars. This is misleading, since dissimilar couples (such as Sirius, Procyon, η Cassiopeiae, ζ Herculis, and 85 Pegasi) are found to be very common near the sun. If these systems were situated ten times as far away, they would be unknown, except perhaps for η Cassiopeiae; they would be seen only as single stars.

The 27 percent of faint pairs are those found either from the *Carte du Ciel* or from the photographic research at Palomar on the extreme dwarfs, which are totally invisible because many of them are at the limit of photographic detection (100 times as sensitive as the eye).

Scarcely half the pairs recorded have known proper motions or spectral types. The following table gives the distribution of types:

	Spectral Type							
	0	B	A	F	G	K	M	Peculiar
Percentage of couples	Very rare	8	25	26	22	16	3	Rare

The middle types are the best represented, as is true of stars in general. This shows that double stars are not differentiated from single ones. It should be noted that photographic discoveries increase the number of red M-type dwarfs.

It is instructive to consider the distribution of binaries as a function of their angular separation. The following cumulative table gives the number of known couples in the northern and southern skies for different limits of separation.

| | No. of Known Couples | |
Separation	Northern sky	Southern sky
≤ 0″.25	569	884
< 0″.5	1,929	3,266
< 1″.0	3,457	6,090
< 2″.0	6,192	10,143
< 5″.0	14,856	18,338
All	39,883	29,976

This table was made in 1975 with the help of the magnetic tapes of the central card index at Washington. Of the 569 very close pairs in the northern hemisphere, nearly half were discovered by Paul Muller and the author 10 years ago. In spite of that, the north is behind the south in the harvest of close couples; half of the southern ones were discovered by Rossiter under the beautiful skies of South Africa at Bloemfontein.

Beyond 5 arc-seconds separation, the stars in a pair are considered widely separated, since at the mean distance of these double stars (300–400 light-years) that separation corresponds to 500 times the distance from the earth to the sun. In these circumstances, the orbital motion is too slow to be detectable; one revolution requires thousands of years. The table shows that more than half the known couples are in this category, made more abundant in the north by the discoveries of Luyten at Palomar.

23 Historical Introduction

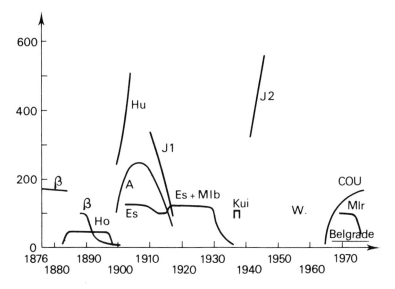

Figure 1.3 A century of surveying double stars to the north of −31°. The ordinate shows the average number of discoveries each year.
A: R. G. Aitken, 3,100 pairs, with refractors of apertures 30 cm and 91 cm at Lick Observatory.
β: S. W. Burnham, 1,336 couples, with various instruments, to which should be added 150 pairs of large common proper motion that were discovered separately.
Belgrade: About 100 pairs discovered since 1966 by P. M. Djurkovic, D. M. Olevic, and G. M. Popovic with a refractor of aperture 66 cm.
Cou: P. Couteau, observations at Nice with refractors of apertures 38 cm (145 pairs), 50 cm (1,400 pairs), and 74 cm (20 pairs).
Es: T. E. Espin (later with Milburn), 2,700 pairs, with reflectors of apertures 40 and 60 cm at Tow Law, England.
Ho: G. W. Hough, 622 pairs with refractor of aperture 47 cm at Dearborn Observatory.
Hu: W. J. Hussey, 1,340 pairs, with refractors of apertures 30 and 91 cm at Lick.
J$_1$: R. Jonckheere, 1,358 pairs; 1,010 of them at Lille with a refractor of aperture 33 cm and 252 at Greenwich with a 71-cm refractor.
J$_2$: R. Jonckheere, 1,942 pairs; 495 of them with the 26-cm refractor at Marseille, 1,368 with the 80-cm reflector at Marseille, and 88 with the 38-cm refractor at Nice.
Kui: G. P. Kuiper, 117 pairs with the refractor of 91-cm aperture at Lick.
Mlr: P. Muller, four discoveries with the 91-cm refractor at Lick and 561 with the 50-cm refractor at Nice.

24 Historical Introduction

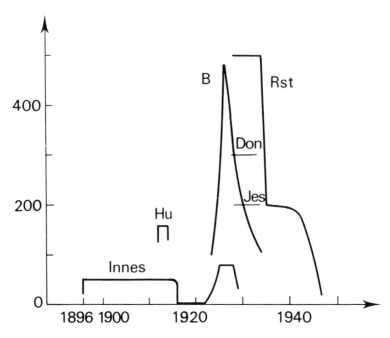

Figure 1.4 Half a century of surveying double stars in the southern hemisphere. The ordinate shows the average number of discoveries each year.
B: W. H. van den Bos, 2,800 pairs at Johannesburg with a refractor of aperture 69 cm.
Hu: W. J. Hussey, 300 pairs with the 43-cm refractor at Johannesburg.
Innes: 450 pairs from 1896 to 1903 with the 45-cm refractor at the Cape of Good Hope, then 1,200 pairs with the large refractor at Johannesburg.
Rst, Don, Jes: R. A. Rossiter, H. F. Donner, M. K. Jessup, 7,368 pairs, of which 5,534 were found by Rossiter.

25 Historical Introduction

Figure 1.5 The 50-cm refractor of Nice Observatory (focal length 7.5 m). Objective by A. Couder, ground by J. Texereau and brought into service in 1967. Nice Observatory photograph.

26 Historical Introduction

Figure 1.6 The 74-cm objective of the large refractor of Nice Observatory about to be mounted at the end of the 18-m-long tube. The scale is evident from the person at the extreme left. Nice Observatory photograph.

BIBLIOGRAPHY

Aitken, R. G. *The Binary Stars.* New York: Dover, 1935.

Baize, P. "L'Astronomie des étoiles doubles." *Bull. Société astronomique de France* 44 (1930): 269.

———. "Catalogue de 240 couples visuels d'étoiles naines rouges." *Journal des Observateurs* 49, no. 1 (1966): 1.

van den Bos, W. H. "Surveys and Observations of Visual Double Stars." In *Basic Astronomical Data.* University of Chicago Press, 1963.

Burnham, S. W. Measures of Proper Motion Stars Made with the 40-inch Refractor of the Yerkes Observatory in the Years 1907–1912. Carnegie Institute of Washington, 1913.

Crossley, E., J. Gledhill, and J. M. Wilson. *Handbook of Double Stars.* New York: Macmillan, 1879.

Flammarion, C. *Catalogue des étoiles doubles et multiples en mouvement relatif certain.* Paris: Gauthier-Villars, 1878.

Heintz, W. D. *Doppelsterne.* Munich: Goldmann, 1971. (English translation: *Double Stars.* Dordrecht: Reidel, 1978.)

Innes, R. T. A. "Reference Catalogue of Southern Double Stars." *Annals, Royal Observatory, Cape of Good Hope* 2 (1899): part II.

Jonckheere, R. "Catalogue and Measures of Double Stars Discovered Visually from 1905 to 1916 Within 105° of the North Pole and Under 5" separation." *Monthly Notices of the Royal Astronomical Society* 61 (1917).

———. Catalogue général de 3,350 étoiles doubles de faible éclat observées de 1906 a 1962. Marseille Observatory, 1962.

Luyten, W. J. "Proper Motion Survey." In K. Aa. Strand (ed.), *Basic Astronomical Data.* University of Chicago Press, 1963.

———. Double Stars with Common Proper Motion. *Astronomical Publications, University of Minnesota,* no. 29 (1972).

Vyssotsky, A. N. "Dwarf M Stars Found Spectrophotometrically: First List." *Astrophysical Journal* 97 (1943): 381.

Vyssotsky, A. N. and B. A. Mateer. ". . . Third List." *Astrophysical Journal* 116 (1952): 21.

Vyssotsky, A. N. ". . . Fourth List." *Astronomical Journal* 61 (1952): 201.

———. ". . . Supplement." *Astronomical Journal* 63 (1958): 211.

Walther, M. E., A. N. Vyssotsky, E. M. Janssen, and W. J. Miller. ". . . Second List." *Astrophysical Journal* 104 (1946): 239.

Webb Society. *Observers' Handbook,* Vol. I: *Double Stars.* London: Mizar, 1975.

2
OPTICAL CONCEPTS USEFUL TO DOUBLE-STAR OBSERVERS

THE MOST FREQUENTLY USED INSTRUMENT: THE REFRACTOR

The measurement of double stars is, above all, an observation at high resolution. It requires instruments that can form theoretical diffraction images that can be examined with adequate magnification. A century and a half passed before it was known how to make objectives for a refractor that would give a good image on the optical axis. The large refractors of the end of the last century can well bear comparison, in the matter of high resolution, with the best modern reflectors. That is one reason why observers of double stars, even today, mainly use refractors. There is another reason: Reflectors are used all the time by teams that relieve each other after short observing runs. This procedure does not allow the observer of double stars to work very fruitfully. He must have his instrument available at the precise, unforeseeable instant when the images are good, and he cannot make observations when the images are of poor quality, even under a very clear sky. On the other hand, refractors are slower than reflectors, more selective, and less adaptable to auxiliary instruments; they are no longer used for spectroscopy and photoelectric photometry, but are reserved for the work of astrometry, of which the measurement of double stars is an important part.

Many of the concepts discussed in this chapter are expounded in greater detail in the work *Lunettes et télescopes* [Refracting and Reflecting Telescopes] by Danjon and Couder (1935). We shall emphasize here only what is fundamental for observers of double stars.

MAGNIFICATIONS: THE EXIT PUPIL

The objective forms an image on a portion of a surface called the *focal plane*. The sharpness of this image depends on the diameter of the objective. The image is examined with a lens of several components, called the *eyepiece,* which can be shown schematically, like the objective itself, as a thin lens (fig. 2.1). As a whole, the optics are *afocal*; that is to say, the beam comes out of the eyepiece in parallel rays. An object subtending an angle α at the objective forms an image at infinity subtending an angle β, seen through the eyepiece. The magnification is the ratio β/α; in other words, the ratio of the focal lengths, F and f, of the objective and the eyepiece:

$M = F/f.$ (2.1)

Nearsighted and farsighted people, whose eyes cannot accommodate to infinity, cannot observe with a strictly afocal instrument, but the difference is always small. Nearsighted people have an advantage; if the instrument is focused for them, they see with a greater magnification. Both nearsighted and farsighted people can

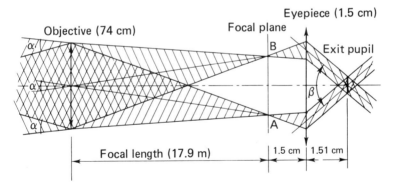

Figure 2.1 Geometry of image formation and magnification in an astronomical refractor. The objective forms an image AB, in the focal plane, of the components of a double star. The eyepiece enables one to see AB subtending an angle β, which is greater than the original angular separation α. For clarity, the diagram is not to scale. The actual quantities for the star Castor and the large refractor at Nice are the following: $\alpha = 2\rlap{.}{''}20$, $\beta = 44'$, $AB = 0.192$ mm, magnification $= \beta/\alpha = 1,200$, diameter of exit pupil $= 0.63$ mm.

have a normal view through an instrument, because the very fine emergent beam has a diameter much smaller than that of the pupil of the eye, so the latter operates at a large focal ratio, minimizing the effects of the faults in the eye lens. That is why the focus of a telescope generally need not be changed for a group of visitors, even though some of them may not be wearing their usual glasses.

The image of the objective formed by the eyepiece is called the *exit pupil*. It is situated practically in the focal plane of the eyepiece. Rays that pass through the objective also pass through its conjugate, the exit pupil. This is where an observer should place his eye to receive the whole of the beam that falls on the objective and is collected by the eyepiece. Inevitably, the beginner puts his eye too far away, thus stopping down the useful beam with his iris and losing a part of the field.

The ratio of the diameters of the objective and the exit pupil is equal to the magnification:

$$M = D/d = F/f. \tag{2.1'}$$

The brightness of the exit pupil is M^2.

According to equation (2.1') there is a minimum magnification below which the exit pupil is bigger than the pupil of the eye at its maximum aperture; this is called the *normal magnification*. Since the pupil reaches a diameter of 0.6 cm (or up to 0.8 cm for some people), this magnification is $M_n = D/0.6$. This is the magnification that the eye of a giant with a pupil of aperture D would have.

If the focal ratio of the objective, $m = F/D$, is substituted, the focal length of the eyepiece that gives the normal magnification will be $f = 0.6$ m, which corresponds to about 10 cm for the instruments used for visual observation.

To observe the focal plane without an eyepiece, the eye should be placed at a distance $\Delta = 0.6$ m from the plane so that the whole beam may enter the pupil. Since Δ cannot be less than the minimum distance for distinct vision, which is 25 cm for normal sight, the focal ratio of the objective would have to be greater than 40. No existing instrument, therefore, allows observation without an eyepiece, but this was not true for the long refractors of the seventeenth century, whose focal ratios reached 290. Those refractors of the time of Molière had to be that long in order that the chromatism of their simple lenses might be reduced to acceptable values.

IMAGE FORMATION BY A CIRCULAR OBJECTIVE; RESOLVING POWER; RESOLVING MAGNIFICATION; USEFUL MAGNIFICATION

For a point source of light, the illumination at a distance n from the geometric center of an image formed by a circular objective is

$$I = A^2 \left(\frac{2J_1(n)}{n}\right)^2 = A^2 \cdot \phi^2(n), \tag{2.2}$$

where A is a constant that depends on the brightness of the star and on the instrument, and $J_1(n)$ is a first-order Bessel function. The distance n can be transformed into values of d [expressed in microns (μm)] by

$$d = 0.321 \lambda mn, \tag{2.3}$$

where λ is the wavelength of the light. The whole set of these illuminations is called the *diffraction image,* or *Airy figure.* It is composed of a central spot (the Airy disk) surrounded by a succession of alternately dark and bright rings. The linear radius of the first dark ring can be written

$$r = 1.22 \lambda F/d = 1.22 \lambda m, \tag{2.4}$$

so the angular radius, a, can be expressed in radians by

$$a = r/F = 1.22 \lambda /D. \tag{2.4'}$$

Since visual wavelengths are effectively centered on 0.55 μm, we have

$$r = 0.67m \text{ (in } \mu\text{m)}$$

and

$$a = 14/D \text{ (in arc-seconds)},$$

when D is expressed in centimeters.

Some remarks are necessary. The symbols r and a denote the same quantity, namely the size of the first dark ring of the diffraction pattern, but r is a linear measure and a an angular one. Also, r is independent of the aperture, but not of the focal ratio; the converse is true for a. The linear radius is about 10 µm in astronomical refractors corrected for yellow light. At the prime focus of large reflectors, of which the focal ratio is close to 3, this radius varies from 2.5 µm in the red to 1.5 µm in the blue. This does not mean that refractors show less chromatic aberration than reflectors, which is obviously not true. In a refractor, however, the light that actually contributes to the diffraction disk is limited to a narrow range of wavelength, about 200 Ångstroms. The other wavelengths are dispersed around the theoretical image and form a large violet halo, not very noticeable visually, which is mainly a mixture of red and blue and is known as the *secondary spectrum.* Refractors give more nearly pure images with clearer rings. The Cassegrain arrangement increases the focal ratio of the mirrors of a reflecting telescope, as well as the size of the Airy disk, but it does not eliminate the dispersion; thus, there is no stable value for the focal length. These considerations are important for the observation of double stars. In a good refractor one can see, in good weather, fine pure images almost like the theoretical ones, with a dark violet halo around the brightest sources; the rings are always clear-cut and do not look colored.

We must now place some emphasis on the fundamental difference between photography and the eye. Photography records linear dimensions; the eye sees angles. In photographing double stars, it is advantageous to choose not only a large focal ratio, to give as large a diffraction pattern as possible, but also a large aperture, since that conformation corresponds to small angles. Visual observation requires only a large aperture combined with fairly powerful eyepieces. Now, although diffraction patterns are easy to see, they are difficult to photograph because they are small. This is the great advantage of visual observations over photographic ones and the reason why, despite all our technical progress, it is very difficult to obtain good photographs of the components of ζ Cancri or ω Leonis, which are easy to see. Worse still, we had to wait until 1971 to discover the companion of θ Coronae Borealis, which is of magnitude 7 and half an arc-second away. It is sufficient to look at this star with a good refractor of 30 cm aperture to see this companion, which still has not been detected photographically with large reflectors. Unfortunately, in our time the sky is not looked at but photographed. The result is that precious information that photography cannot give us is lost. Is it not curious that we can detect stars of magnitude 23, lost in the depths of the Galaxy, yet cannot photograph a star of magnitude 7 because it is half an arc-second from another of magnitude 4?

The *resolving power* is, by definition, 85 percent of the angular size of the first dark ring; that is, in arc-seconds,

$$p = 0.85a = 12/D. \tag{2.5}$$

Thus, an objective of 10 cm aperture has a resolving power of 1.2 arc-seconds (1".2), which would correspond to two Airy disks, not separated but discernible as a flattened figure-eight. The resolving power is an arbitrary notion that does not correspond to any instrumental limit.

The *resolving magnification* is the magnification that makes the size of the first ring equal to the limit of visual acuity, supposed to be 1 arc-minute. We have, then,

$$pM_r = 60'', M_r = R \text{ mm}. \tag{2.6}$$

This means that the resolving magnification is numerically equal to the radius of the aperture R, expressed in millimeters. It is also equal to three times the normal magnification. The resolving magnification just makes the diffraction image visible. To see it in detail (for example, while observing a close binary), one must use a higher magnification, called the *useful magnification,* which is usually three to four times and may be up to five times the resolving magnification. With still higher magnifications, the edges of the image are badly defined and the eye cannot transmit information. Measures of double stars ought not to be made with magnifications less than twice the resolving magnification.

This summary of basic concepts will enable us to study the structure of the image of a close double star.

IMAGE STRUCTURE OF A CLOSE DOUBLE STAR

In this context, a close double star is one whose components are separated by an amount less than or equal to the radius of the first dark diffraction ring. The components of a couple are completely resolved if their separation is greater than the diameter of the first dark ring. The visual impression is different, because the space between the stars appears dark even if the separation is less than this diameter. At what point does this impression cease? Can it be related mathematically to the theory of diffraction? We stress this point because opinions do not always coincide with the limits of resolution of instruments. Danjon and Couder defined the limit of resolution in their work cited above, with supporting graphs, and gave the limit as 85 percent of the radius of the false disk. But visual observers have measured much closer double stars with relative ease. The limit of the resolution can, therefore, be exceeded; but by how much? The question is often posed, and the answers are varied. We shall analyze this problem of "vanishing duplicity" and relate it to the visual impressions.

Consider the focal plane in which are two rectangular axes x and y (fig. 2.2). Component A is at the origin, and component B is on the x axis at distance x_0 from A. Recall equation (2.2), which describes the diffraction pattern from a point source. For a double star whose components are not resolved, the intensities add and the illumination at a point $P(x, y)$ in the focal plane is

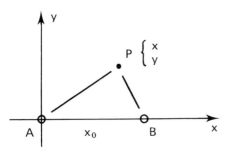

Figure 2.2 Illumination of a point in the focal plane. The two components A and B each contribute to the illumination at a point P.

$$I(x, y) = A^2\phi^2(x^2 + y^2)^{1/2} + B^2\phi^2[(x - x_0)^2 + y^2]^{1/2}. \tag{2.7}$$

The equations of the isophotes can be written

$$I(x, y) = \text{constant}. \tag{2.8}$$

Detailed study of the curves determined by the equation of the isophotes enables us to interpret the visual impressions given by a double star whose components are closer than the limit of resolution.

Note first that the luminous intensity in the Airy disk has a very pronounced maximum at the center (fig. 2.3): a peak of light called the *photocenter*. A very close binary shows a double or elongated photocenter, which contributes as much as the overall appearance of the disk to the recognition of duplicity. We have gathered in table 2.1 the results of calculations on, and the interpretation of, a theoretical close couple with equal components. The figures are based

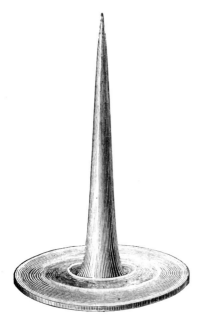

Figure 2.3 The solid of diffraction. Heights are proportional to the illumination. Note the intensity of the central peak.

Useful Optical Concepts

Table 2.1

$d(r)$ (Image Theory)[a]	$d(r)$ (Photocenters)[b]	I_M/I_m[c]	Image Elongation[d]	Elongation of 0.5 Isophote[e]	d (50 cm)[f]	d (Photocenters) (50 cm)[g]	Appearance of Image[h]
1	1	0.748	1.47		0″280	0″280	Separated
0.95	0.91	0.827	1.45		0″266	0″255	Tangential
0.90	0.80	0.902	1.425	2.08	0″252	0″224	Figure-eight
0.85	0.64	0.961	1.40	2.00	0″238	0″179	Flattened eight
0.80	0.35	0.995	1.38	1.90	0″224	0″098	Narrow rod
0.75	0	1		1.77	0″210	0	Rod
0.70			1.35	1.64	0″196		Rod
0.60			1.30	1.42	0″168		Olive
0.50			1.25	1.26	0″140		Slightly oval

a. Distance of the components as a function of the radius of the Airy disk.
b. Distance of the photocenters.
c. Relative intensity of the middle of the image.
d. Elongation defined by the ratio of the major to the minor axis.
e. Elongation of the isophote 0.5 (corresponding to half the maximum intensity).
f. Corresponding separations for the 50-cm refractor at Nice.
g. Separations of the photocenters for the 50-cm refractor at Nice.
h. Appearance of the image with adequate magnification.

on isophotes determined at the Center for the Reduction of Astronomical Plates (CDCA) at the Nice Observatory. This table is rich in information, and some comments will be useful. For a separation equal to the radius of the first dark ring, r, the relative intensity at the center of the double image is about 0.75. This is sufficient to give the illusion of complete resolution; that is to say, of a thin dark space between the components. The effect is particularly clear for binaries that are not too bright, whose diffraction rings are scarcely visible. For smaller separations, down to $0.8r$, two photocenters can be distinguished. It is important to note that they do not coincide with the geometric images of the components; they are closer together. The deviation, zero for a separation equal to the radius of the first dark ring, grows quickly, as the table indicates. Thus, for a separation of $0.85r$, corresponding to the resolving power, the photocenters are $0.64r$ from each other, separated by an "intensity hollow" of 0.96 that is still discernible if the components are fairly luminous.

It follows that filar micrometers, the principle of which is based on the bisection of maxima of light by fine threads, cannot give exact measurements for unresolved double stars; the same is true for double-image micrometers. Experience shows that the best micrometer is the diffraction image itself. An observer accustomed to his objective and having to his credit years of practice and thousands of measures cannot go far wrong in estimating the separation according to the appearance of the image. He should, however, submit his experience to calculation in order to avoid systematic errors. To be sure, novices are not encouraged to begin by observing close pairs; but it must be recognized that the dynamically most interesting double stars, and those that remain to be discovered, are mostly binaries whose components are unresolvable even with large instruments. Our knowledge of masses automatically depends on the care that we bring to the interpretation of the angular distances of stars whose images impinge upon each other.

Figure 2.4 shows the isophotes of the center of the Airy disk for three theoretical couples, of separations $0.9r$ (a little above the limit of resolution), equal to the limit, and a little below ($0.79r$). Note how obvious the photocenters are at the first two separations, and the considerable elongation of the center of the disk at the third. The

Figure 2.4 Isophotes of the center of the Airy disk of an unresolved double star with equal components (CDCA photograph). Note the considerable changes in the theoretical appearance of the disk between 0.9r and 0.8r. Upper photo: Separation 0.9r (0″.504 for an aperture of 25 cm); the photocenters are well separated. Middle photo: Separation 0.85r (0″.476 for an aperture of 25 cm); the photocenters are still distinct. Lower photo: Separation 0.79r (0″.442 for an aperture of 25 cm); only an appreciable elongation of the center can be seen.

appearance of the center of the image reveals more than that of the whole image. At first glance, the observer will note that the image of a very close double star is not single by the look of the center of the disk, even before he distinguishes the elongation of the whole image.

For separations less than 0.79r, the image of two stars presents only one photocenter. The whole image is elongated. This elongation can be defined by the ratio of the major axis to the minor axis of the zero-intensity isophote. This is not necessarily the elongation of the image as it strikes the eye, since near the zero isophote the intensity varies little. Intensity varies most rapidly at the level of the isophote 0.5; experience shows that it is the shape of this isophote that most faithfully reproduces the elongation actually seen. This is given in the fifth column of table 2.1. Note that its elongation is always greater than that of the entire disk—it is 1.42 for a separation of 0.6r; that is, the same as the elongation of the entire disk at the limit of resolution. This enables us to understand how observers can see and successfully measure couples much closer than the limit of resolution. The isophotes all tend to the same elongation for a separation of 0.5r, which marks the limit of detection of duplicity.

39 Useful Optical Concepts

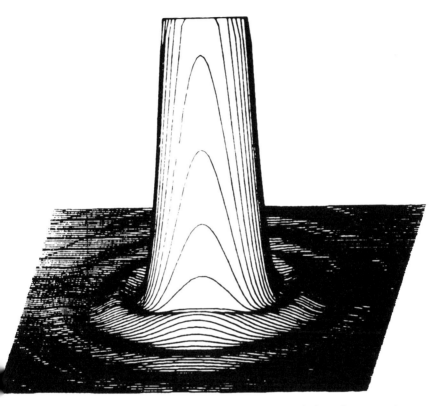

Figure 2.5 Isophotes of an unresolved double star consisting of two equal components. Note the two symmetrical nodes of light in the first ring. CDCA photograph.

40 Useful Optical Concepts

The sixth and seventh columns of table 2.1 give the angular separations of stars and photocenters for the Nice 50-cm refractor. The last column, the fruit of several decades of observing close double stars, gives a description of the appearance of the image that is valid for stars between magnitudes 6 and 9.5, with magnifications up to four times the resolving magnification.

Before the close of this section, the difficulty of observing a close double star must be stressed. The movement and diffusion of images, caused by atmospheric turbulence, so disturbs the Airy disk that its examination and interpretation are often impossible and always difficult for any instrument of aperture greater than about 30 cm. A casual observer will see only a boiling disk where an experienced eye recognizes the characteristic appearance of a double star. This is why amateurs are encouraged to train themselves with instruments of modest size. An aperture of 10 cm is quite suitable; the images are stable and the interpretation of the luminous disk much easier. There are lists of pairs that will provide the conditions of table 2.1 for observers. The examination of pairs such as ζ Bootis at 1″.13, ζ Cancri at 0″.9, ξ Scorpii at 1″.24, and many others gives the amateur the means of testing his instruments and his aptitude for observation. If he has trained well with a modest aperture, the amateur who is entrusted with a powerful instrument will be able to analyze with it the critical and capricious images that these optical colossi show, and thus convey precious information to those who calculate orbits and stellar masses.

It remains to recall what is meant by the intensity of an image formed by an instrument, a parameter that also limits the possibilities of observation but is less important here than in the observation of planets and extended sources in general.

IMAGE INTENSITY

The image intensity is the gain in light given by an instrument relative to the naked eye. It is necessary to distinguish between point sources (stars) and extended sources.

In the stellar case, the image intensity is the ratio of the luminous flux collected by the instrument to that which enters the pupil. If D

41 Useful Optical Concepts

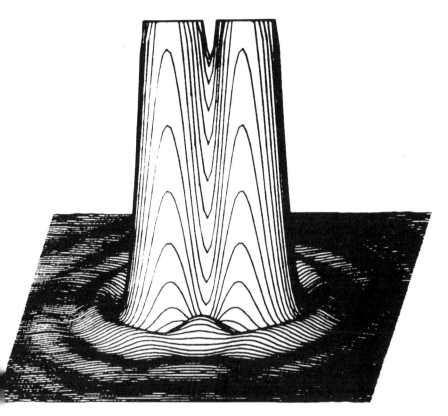

Figure 2.6 Isophotes of a double star consisting of two equal components whose separation equals the resolving power. There are prominent nodes in both the first and second rings. The contrast between the middle of the disk and the images of the components is clearly visible. CDCA photograph.

and δ are the apertures of the instrument and the pupil, we have

$$C = (D/\delta)^2 = M_n^2, \tag{2.9}$$

where C is the image intensity. For stars, the image intensity is always greater than unity and equal to the square of the normal magnification. Thus, with a telescope of 1 m aperture one can see stars 27,800 times as faint as those that can be seen with the naked eye. The limiting visual magnitude can be written (see Danjon and Couder)

$$m = 7.1 + 5\log D.$$

The image intensity is independent of the magnification as long as the latter is between M_n and $15M_n$; beyond that the image can no longer be considered as a point.

For extended sources it is necessary to distinguish between visual and photographic observation. In visual observation the light collected through the aperture is distributed over an image magnified M times, and the image intensity can be written

$$\Gamma = C/M^2 = (M_n/M)^2. \tag{2.10}$$

This image intensity, then, is less than unity, unlike that of stars. If the normal magnification is used, the image intensity (of an extended source) is equal to unity. Appreciably extended objects, seen through a refractor or reflector, are less brilliant than to the naked eye, or at most of equal brightness.

These considerations are important when one looks at the sky background, comets, or nebulae. Through a telescope, the sky background is always darker than for the naked eye and becomes darker as magnification is increased. A comet or a nebula is easily visible only with the normal magnification, but even then not all the detail that the objective can show will be visible. To explore the nucleus of a comet, for example, the magnification should be increased, but then the coma and the tail of the comet will no longer be visible. For a nebula, where there is little contrast in the details, visual observation is strongly deceiving.

Terrestrial observations are most pleasing with the normal magnification, or one a little higher; otherwise the brilliance is lost. At night, silhouettes can be seen only with this magnification. In this

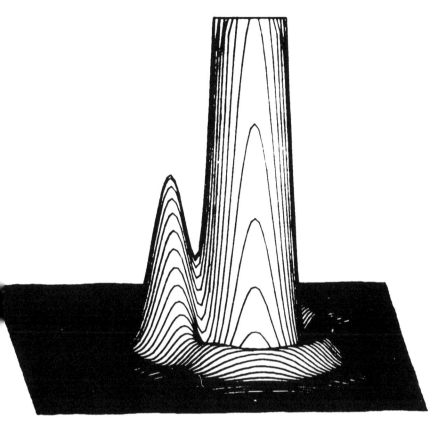

Figure 2.7 Isophotes of a double star with unequal components. CDCA photograph.

case, contrasts are not seen any better than with the naked eye, but the image is magnified and therefore more detail is seen. An error often committed by beginners is to increase the magnification of their instrument, the better to see a landscape.

The observation of planets requires at the same time good resolution and the highest possible contrast between marks on the disk. These two conditions are contradictory. A planetary disk will therefore be seen very differently according to the eyepiece used. Lunar details strongly contrasted can bear high magnification, while the planets Mars and Saturn, which show little contrast, cannot. The difference of contrast given by two instruments can be demonstrated by supplying the same magnification to an instrument and its finder. Through the two objectives the planet subtends the same angle; but the large instrument gives more contrast than the other, through which the disk seems dull and washed out.

The photographic observation of extended surfaces will not be emphasized. Suffice it to say that the image intensity Γ_p is proportional to the area of the objective and inversely proportional to the area of the image. The difference from visual observations is that one works with images, not angles. Thus, we can write

$$\Gamma_p = (D/F)^2 = 1/m^2. \tag{2.11}$$

In photographic observation the image intensity is equal to the reciprocal of the square of the focal ratio. Two objectives of the same ratio give the same image intensity; it takes them the same exposure time to photograph an object of appreciable extent, or the sky background. But the larger of the two gives a larger image. Two plates of the same part of the sky, taken with two objectives of the same size but different focal ratios, will not be similar. The instrument of shorter focal length will show the nebulae, but is more sensitive to the illumination of the sky background; the other will show finer structure, but at the expense of contrast. Table 2.2 summarizes the preceding discussion.

Useful Optical Concepts

Table 2.2

Aperture (cm)	5	10	50	100	150
Resolving power	$2''\!.4$	$1''\!.2$	$0''\!.24$	$0''\!.12$	$0''\!.08$
Visual limiting magnitude	10.6	12.1	15.6	17.1	18.0
Normal magnification	8	17	83	167	250
Field[a]	$5°$	$2°20'$	$29'$	$14'$	$10'$
Resolving magnification	24	51	250	500	750
Field[a]	$1°\!.7$	$47'$	$10'$	$4'\!.8$	$3'\!.2$
Maximum magnification	125	250	1,250	2,500	3,750
Field[a]	$20'$	$9'$	$1'\!.9$	$1'$	$36''$

a. True field (i.e., maximum value of α in figure 2.1) for an assumed apparent field (maximum value of β) of $40°$.

BIBLIOGRAPHY

Bruhat, G. *Optique,* fifth edition revised and completed by A. Kastler. Paris: Masson, 1959.

Danjon, A., and A. Couder. *Lunettes et télescopes.* Paris: Editions de la Revue d'Optique théorique et expérimentale, 1935.

Francon, M. *Optique: Formation et traitement des images.* Paris: Masson, 1972.

Marechal, A. *Diffraction. Structure des images.* Paris: Masson, 1970.

3
MEASURING INSTRUMENTS

VISUAL MEASURES

Principles of Double-Star Measurement

The measurement of a double star consists of the determination of the polar coordinates of the companion with respect to the primary star, which is taken as the origin. The origin for angles is the direction of celestial north. This does not correspond to anything observable in the field of an instrument. It would have been more logical to take as origin the east-west trajectory of an equatorial star in its daily motion, as Sir William Herschel did in the eighteenth century. This diurnal motion can be carefully determined by watching a star drift along the micrometer thread several times. For equatorial mountings, this motion should be the same as that of the telescope; it can be determined once and for all, even though it should be checked from time to time. The north-south direction is perpendicular to that of the diurnal motion.

The field of the instrument is divided into four quadrants. The first goes from north to east (0° to 90°), the second from east to south (90° to 180°), the third from south to west (180° to 270°), and the fourth from west to north (270° to 360°), as shown in figure 3.1. The angle between the line joining the components of a pair and the direction of celestial north is called the *position angle* (fig. 3.2) and is usually denoted by θ. The *separation* or *distance* of the components, measured in arc-seconds, is usually denoted by ρ. A complete measure comprises the epoch (in years and fractions of a year), the position angle θ, and the separation ρ. The purpose of the micrometer is to make the measurement of position angle and separation possible.

47 Measuring Instruments

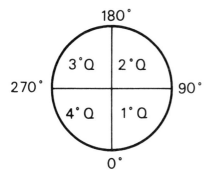

Figure 3.1 Quadrants in a telescopic field.

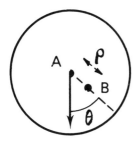

Figure 3.2 The principle of measurement of a double star. A and B are the components. The arrow indicates north. θ and ρ are the quantities to be measured.

Precautions to Take in Fixing the Origin for Angles

Since the diurnal motions of stars are not along great circles of a sphere, unless a star is situated on the celestial equator it will not follow a straight line in the field of the instrument. The diurnal motion cannot be rigorously represented by a stretched thread. The difference is small, and imperceptible a long way from the pole, near which a micrometer should not be used.

It is important to make sure that the polar axis is correctly oriented. A displacement of even a few arc-minutes will have an appreciable effect on the position angles. As double stars ought to be observed near their culmination, the east-west displacement of the polar axis plays an important role. This is easy to recognize and correct, because refraction does not interfere. The way to go about it is as follows. Set on a star close to the equator and the meridian, bisect the image with one of the east-west threads of the micrometer, and set the drive going. If the polar axis is well adjusted, the star will stay on the thread; but if, after some minutes, the star moves up or down, then the axis is displaced. In effect, the movement of the telescope traces a circle whose pole is the extension of the polar axis, while the star traces a circle centered on the real pole. This can be expressed in equations; but there is no point in doing so, since all we are interested in is getting rid of the displacement, not knowing its value. If the star veers towards the north, the axis is displaced to the west and the corresponding screws on the column of the instrument should be adjusted. Of course, the complete adjustment of an equatorial mount includes several other operations, which are described in detail by Arend (1951). Theoretically, an instrument once set up should not get out of adjustment, but subsidence of the ground and microseisms can change the orientation of the axes slightly over the course of decades.

Micrometers

Many kinds of micrometers have been conceived. We shall study here only those that have been used to measure double stars. They can be reduced to four kinds: filar micrometers, comparison-star micrometers, interference micrometers, and double-image micrometers.

The Filar Micrometer

The principle of this instrument goes back to Auzout in the eighteenth century. Mechanical difficulties of construction delayed its practical application until the time of Wilhelm Struve in the nineteenth century. Before this, only meridian circles were available to determine the positions of objects, but they did not lend themselves well to the measurement of small angles. The first truly useful observations of double stars were made about a century and a half ago.

A filar micrometer is composed essentially of a metallic frame that moves along two parallel ways forming the long sides of a rectangle

49 Measuring Instruments

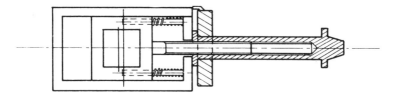

Figure 3.3 Schematic diagram of a filar micrometer. Note the two parallel wires, one of which is connected to the moving carriage controlled by a screw and a graduated drum. From Danjon and Couder, *Lunettes et télescopes*.

(fig. 3.3). The surfaces of the frame and the ways are coplanar; they must work in the focal plane. A cross of very fine threads is held in the middle of the sides of the movable frame, and another very fine thread is held between the ways in such a manner that one of the threads on the frame is exactly parallel to it. The frame can move along the ways; its position is determined by a comparator or a micrometer screw attached to a graduated drum. The whole instrument turns smoothly about the optical axis. The threads come from spiders' webs or silkworms; nylon or quartz threads, not affected by moisture or aging, are now made industrially. The diameter of these threads is of the order of 7 μm. In some recent micrometers the screw serves only to move the frame, whose position is determined by an electronic gauge connected to a counter that displays and prints the displacements in microns.

Some observers, like Jonckheere, prefer a micrometer with oblique threads. In this arrangement (fig. 3.4), the moving frame carries two threads that make a small angle with each other and intersect at the center of the frame. The ways carry a thread perpendicular to the bisector of the small angle. Measurement is made by keeping the image of a stellar couple on the fixed thread and maneuvering the movable frame until the sides of the angle are tangent to the images. The operation is repeated on each side of the intersection; thus, at the same time one measures separation and position angle.

That is the very simple principle of the filar micrometer. It is difficult to make one—partly because all the threads ought, by definition, to be in the same plane (which is mechanically impossible),

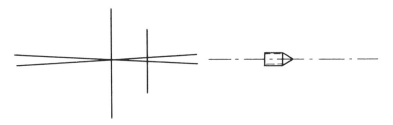

Figure 3.4 The principle of a filar micrometer with oblique wires. The two oblique wires, connected to the moving carriage, are carried in front of a wire that is perpendicular to the axis of the screw. From Jonckheere, *Un micromètre à fils obliques*.

and partly because no play can be permitted in the sliding of the moving frame or in the rotation of the screw. The threads are set in the same plane by a fine adjustment of the inclination of the ways with respect to the plane of the movable frame. The depth of focus of the eyepiece sets a limit on the displacement of the planes. With an eyepiece of focal length 6 mm, the tolerance is about 50 μm. The play of the frame is compensated by springs; the backlash of the screw is no longer important when the measurement is made by an electronic gauge. The threads are made visible in the eyepiece by illuminating either the field or the threads themselves. The field is illuminated simply by projecting light on the objective from a lamp of adjustable power situated not far from the eyepiece. Adjustable illumination of the threads is by a small luminous source in the plane of the threads in the base of the micrometer.

The position angle is measured by visual estimation of parallelism between the fixed thread and the line between the components; the measurement is made several times by turning the micrometer about the optical axis. To measure the separation, each component is bisected by the parallel threads, after they have been oriented perpendicularly to the line joining the components. The bisection is repeated, exchanging the positions of the threads with the help of the graduated drum. The difference of the drum readings is proportional to the separation in arc-seconds.

Precautions for Obtaining Good Measures with the Filar Micrometer

Setting the micrometer threads in the focal plane This eliminates two sources of systematic error: the effect of parallax introduced by the nonsuperposition of the images and the threads, and the error in the focal length of the instrument. These sources of error are important (especially the first when the images are widely separated), and can reach 10 percent of the measure. The effect of parallax is checked by looking at the image close to the thread and moving the head. If the setting at the focus is incorrect, as it fairly often is in the reticules of finders, the star and the thread will not seem to keep the same relative position. It is then necessary to make sure that the focus is correct by focusing the eyepiece on the threads, and then setting the micrometer on the stellar image. The tolerance in the focusing setting of a big refractor is some tenths of a millimeter.

Calibration of the micrometer screw Until recently, filar micrometers were all provided with micrometer screws. This was illogical in principle, because the micrometer screw both drove the carriage and determined its displacement. These functions are incompatible; a measuring instrument ought to do no work, otherwise it will wear out. Screw errors can be periodic (arising from wear of the pitch), a function of the portion of the screw used (arising from stretching of the springs), or a function of the temperature. Their calibration is difficult; it consists of determining the equivalent of the pitch in arc-seconds. A very good first approximation can be had by dividing the value of the pitch by the focal length expressed in the same units. An approximate value in radians is thus obtained, which is converted to arc-seconds by multiplying by 206,265. The final calibration is made by repeated observation of couples of well-known separation, or, better, by the time that it takes a star to traverse the space between two threads separated by a known number of revolutions of the screw. Good photographs of star fields such as the Pleiades give the focal length and the number of arc-seconds in a micron. Nowadays it is convenient to use a comparator instead of a micrometer screw. There are good instruments available commercially, graduated in tens of microns and easy to read to a micron. The 50-cm and 74-cm refractors at Nice have, besides micrometer

52 Measuring Instruments

Figure 3.5 The front of the 74-cm refractor at Nice. **A:** Case in which the controls for moving and lighting the dome, for the pneumatic clamps in right ascension and declination, for the slow motions, and for the circle and sky-background lights are placed. **B:** Focus control switch. **C:** Filar-micrometer case. **D:** 25-mm eyepiece in position with its focus ring. **E:** Control for micrometer-wire lights. **F:** Control for lighting position-angle circle and drum. **G:** Micrometer-screw drum. **H:** Lights for the position-angle circle and drum. **I:** Drum reader. **J:** Microphone for magnetic tape. **K:** Position-angle circle, and adjustment. **L:** Eyepiece store. **M:** Declination viewer. **N:** Display showing separation of micrometer wires. **O:** Handles for slewing in right ascension and declination. **P:** 25-cm finder. **Q:** Guiding telescope. **R:** Safety switch. Slow-motion and microphone controls are also available on a paddle not shown. Nice Observatory photograph.

screws, comparators and electronic gauges. These instruments can easily be changed, since they are not part of the micrometer. Their use eliminates all the errors arising from the screw.

Illumination of the threads Eyepieces are not achromatic; they have different, although nearly equal, focal lengths for each wavelength. The threads must be illuminated in white light, otherwise the focus will be different for the wire and the star. This error is called lateral chromatic aberration. It is suppressed if the field is illuminated, but then faint stars are difficult to observe. Illumination of the threads is a source of light whose importance depends on the exit pupil of the eyepiece itself, and not just on that of the whole telescope as in the case of starlight. Care should be taken, therefore, to make sure that the exit pupil of the eyepiece does not greatly exceed that of the whole instrument—a condition rarely met with in commercially available eyepieces.

Checking the parallelism between the threads and star alignments During the measurement of a position angle, the observer must always have the line of his eyes either parallel or perpendicular to the alignment of the stars. This is not always easy, because some angles require uncomfortable positions of the head. Some observers use a totally reflecting prism that changes the position angle by a known amount. This loses light, however, and is a complication in the eyepiece mounting; compensating prisms are no longer used much.

Checking the bisection In principle, a measurement consists of bisecting the image of a star by a thread. This operation is difficult because of movements of the mounting, or others of atmospheric origin, which often amount to a perceptible fraction of the distances to be measured. During the measurement, the images ought to be kept in the same region of the field (if possible, close to the center) in order to avoid any large error arising from nonparallelism of the threads. Bisection becomes impossible if the images are not completely separated; diffraction sets the limit. Double-star observers know these difficulties well. It is not easy to measure the separation when atmospheric turbulence combined with diffraction no longer permits the components of the system to be seen distinctly. If turbulence is very weak, a good estimate of the separation can be

54 Measuring Instruments

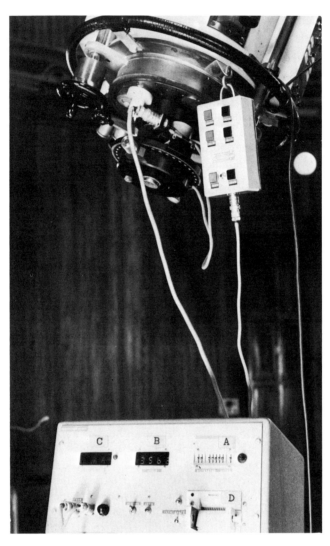

Figure 3.6 (Left) Recording of observations with the 50-cm refractor at Nice. **A:** Star identification. **B:** Position-angle display. **C:** Angular-separation display. **D:** Printer. (Right) Mounting the 74-cm objective, May 29, 1969: J. Texereau putting the lenses in their cell. Nice Observatory photographs.

55 Measuring Instruments

made by the appearance of the Airy disk, but this is no longer a micrometer measurement.

Filar micrometers are used by at least nine out of ten observers, because they have the great advantage that they leave the image formed by the objective intact. But these instruments do not have the impersonality of modern interferometric and double-image micrometers. It takes years of practice to achieve good results with filar micrometers, because bisecting an image is difficult. Measurements of double stars carry the hallmark of the observer, since their quality depends on the care he takes even more than on the power of his instrument. Their advantage over photographic methods is evident in the measurement of images that can be seen but not photographed.

Comparison-Star Micrometers
I will not emphasize these instruments which, to my knowledge, are no longer employed. Baize (1949) described them well. Two artificial stars, whose separation and orientation are both variable and determinable, are compared with the couple to be measured. In the arrangement devised by Davidson and Symms (1931), light from a point source passes through a Wollaston prism and produces two images whose separation is related to the distance between the source and the prism; crossed Nicol prisms can vary the brightness of the image. The whole instrument can turn about an axis, and the images are sent into the telescope by a series of reflections. This ingenious arrangement has not been very fashionable; even its designers have made very little use of it.

By contrast, M. Duruy's micrometer (1937) has been used by him for many years. This is a binocular system: One eye looks through the telescope, the other through a sight at an adjustable image of an artificial double star formed by two white points. In spite of the physiological complexity of the measurement (one eye observes and the other compares), numerous observations have been made, often of objects that are difficult for the 25-cm refractor that Duruy uses.

Interference Micrometers: The Fizeau-Michelson Interferometer
Comparison-star micrometers, like filar micrometers, are difficult to

use when the components are too close to be seen separately. Fizeau had the idea to fit in front of the objective an arrangement of Young's slits that would give interference fringes from a point source. For this purpose a screen is placed in front of the objective, or in front of the focal plane (which is less cumbersome), and two equal apertures are pierced in it. They are arranged symmetrically with respect to the optical axis, and their separation D can be varied. In the focal plane light is received from each aperture. At a given position, the difference of travel from the two slits is a function of the distance to the optical axis; thus fringes are observed. It is easy to show that the space between fringes is equal to λ/D in radians, to be compared with the value $1.22\lambda/D$ of the angular radius of the diffraction disk. By varying the distance between the apertures, and their orientation, one can make the dark fringes from one component of a pair of stars coincide with the bright fringes from the other, provided that the components are fairly bright and not very different in brightness. The fringes disappear; the observer notes the orientation and the separation of the apertures.

This micrometer gives access to separations smaller than the resolving power. It allowed Finsen, in particular, to discover some very close couples, of which one has the shortest known period of a visual double (1.59 years). Other observers, such as H. M. Jeffers, have obtained results no less excellent. The instrument is slow, however, because it intercepts the greater part of the luminous flux that reaches the objective, and the number of fringes further diminishes the contrast. These considerations led Danjon (1937) to devise a faster interference micrometer that produces only one fringe (see figure 3.7).

The Half-Wave Interference Micrometer

A Jamin compensator, formed of two thin glass plates with parallel sides mutually inclined at a small angle, is placed in front of the objective, each plate covering one half of it. Because of the small difference in inclination, light coming out of the plates is out of phase. The difference in pathlength is arranged to be half a wavelength. The image appears as a double luminous spot with a central dark fringe. A diaphragm placed in front of the objective defines the outline. The angular dimensions of the double spot are inversely

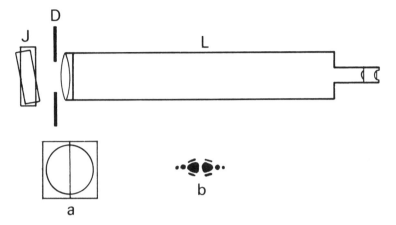

Figure 3.7 The principle of Danjon's microphotometer. A Jamin compensator (**J**) formed of two identical parallel-sided plates is placed in front of the objective (**a**). A diaphragm (**D**) stops down the aperture and changes the appearance of the image. From *Annales de l'Observatoire de Strasbourg*.

proportional to the aperture of the diaphragm. The whole instrument can be turned about the optical axis. To make the diffraction image more nearly pure, the diaphragm is a square.

A double star gives a system of two images with a dark fringe. By orienting the instrument correctly and opening the diaphragm progressively, a pure fringe is observed at first, which shades away and disappears when the double image has the same separation as the couple, that is, when

$$\rho = 0.77\lambda/d, \tag{3.1}$$

where d is the length of the side of the diaphragm. When the objective circumscribes the diaphragm, the separation is at a minimum and we have

$$\rho_{min} = 1.09\lambda/D, \tag{3.1'}$$

where D is the aperture. Thus the resolving power of the objective can almost be reached.

This instrument can serve as a double-image micrometer to measure either wide or close couples. By turning the micrometer and opening the diaphragm a greater or lesser amount, geometrical figures can be formed with the four images coupled two by two in squares, alignments, or lozenges, the regularity of which the eye is particularly good at judging.

The inventor of this micrometer measured hundreds of couples with the 49-cm refractor at Strasbourg, making an important contribution by the precision of his measures. The advantage of this interferometer over that of Fizeau and Michelson is its greater speed and image contrast, thanks to the single fringe. Nevertheless, it intercepts some of the beam reaching the objective; the actual amount depends on the double star being observed. In fact, the greatest merit of Danjon's double-image micrometer was that it led Muller to design another of a new type.

The Muller Double-Image Micrometer

The theory of this instrument, conceived in 1937, was developed in 1947. We shall lay some stress on this instrument because it is fairly widespread, thanks to the ease of its construction. Paul Muller placed two birefringent half-prisms near the focal plane of the objective. These half-prisms, cemented together, have their axes in the plane of the principal section. These axes make an angle of 45° with respect to the entrance and exit faces of the birefringent prism. The result is that a ray incident almost normally on the prism is divided into two rays, ordinary and extraordinary, which emerge parallel at a distance from each other that depends on the lateral displacement of the prism.

It is easy to show that this is so. Let OI be the ordinary ray, and let OME be the extraordinary ray, which is refracted through the angle α at the entrance face, through 2α at the cemented faces, and again through α at the exit face (fig. 3.8). Let the thickness of the double prism, AB, be $2l$. Project M onto OI at H, and let $OA = x$. This quantity measures the displacement of the prism. The angle is small; therefore, to the first order of small quantities we can write

$OM = OH = x \tan\beta = x,$ since $\beta = 45°.$

Then, because M' is the projection of M onto the exit face, we have

$EI = EM' - IM' = EM' - HM.$

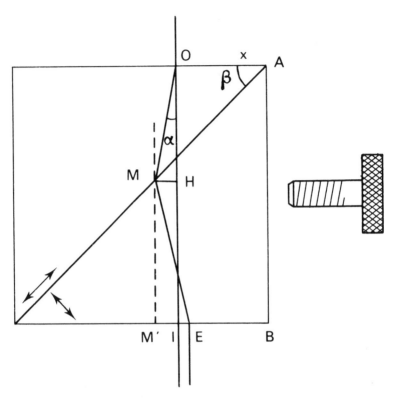

Figure 3.8 The Muller micrometer. Two quartz prisms are cemented with their optical axes perpendicular and parallel to the plane of the cut that makes an angle β with the incident face. The extraordinary light ray travels along D,M,E and exits parallel to the ordinary ray. Their separation is proportional to the displacement $DA = x$, which is controlled by a screw and a graduated drum.

61 Measuring Instruments

Now $HM = \alpha x$ and $EM' = \alpha(2\ell - x)$, since $\beta = 45°$; therefore,

$$EI = 2\ell\alpha - 2\alpha x. \tag{3.2}$$

A rigorous calculation gives

$$EI = 2\ell \tan\alpha - \frac{2x \sin\beta}{\cos(\beta + \alpha)} \sin\alpha. \tag{3.2'}$$

The quantity $\gamma = 2\alpha = 2(n_o - n_e)/n_o$ is a constant, where n_o and n_e are the ordinary and extraordinary indices of refraction of a quartz Muller prism. The reciprocal of this constant is called the *magnification* and has the value 84.0. Note that $EI = 0$ when $x = \ell$. The images coincide in the middle of the incident face of the prism; the separation of the images is proportional to the displacement of the prism from the position in which the images coincide, the factor of proportionality being γ. It is as if the number of revolutions (R) of the screw required to make this displacement has been divided by the magnification. If R_0 is the separation of the images, expressed in revolutions of the screw, we have

$$R_0 = \gamma R. \tag{3.3}$$

The separation of the images is not, however, the same for all stars in the field; it is, therefore, not exactly the same even for the two components of a couple. This is called the *differential micrometric effect*.

To measure a couple, an alignment or a square is formed from the images. Alignment consists of arranging the four images A, B, A_1, and B_1 in the same straight line on each side of the point where the images would coincide, and making them equidistant from each other, either by separating the two pairs of images completely (4d method, figure 3.9) or by alternating them (simple-distance method, figure 3.10). Because of the micrometric effect, it is impossible to

Figure 3.9 Alignment of images (4d method).

Figure 3.10 Alignment of images (1d method).

make the images exactly equidistant, but the deviation is not noticeable to the eye. The resulting error may be corrected by adopting an effective number of screw revolutions:

$$R_1 = R_0(1 - \gamma/2). \tag{3.4}$$

The position angle may be measured with great precision by the alignment method. If the square arrangement is used, the micrometric effect does not play a role, and the separation in screw revolutions remains R_0. The square arrangement (fig. 3.11) is convenient for measuring close couples.

The differential micrometric effect can be removed by using a compensating prism. This is a similar prism, but reversed and fairly small so that it can be put in the mounting of the eyepiece. It can be shown by a simple calculation that, in this case, the stars in the field have very nearly the same separation, which, measured in revolutions, is $R_0 = \gamma R$ for all methods. Nevertheless, a small micrometric effect remains, if one is to be completely rigorous, because of the differential inclination of the beam on the prism. This effect never amounts to a thousandth part of the quantity being measured.

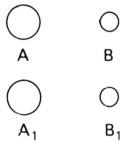

Figure 3.11 Square arrangement of images.

Finally, we note that the eyepiece must be a microscope, since the focal plane passes through the prism and does not allow a sufficient magnification for ordinary eyepieces.

This instrument can easily be constructed for a modest sum, by replacing the wires in the frame of a micrometer with a prism. The field does not exceed an arc-minute, but this is not inconvenient. There are Muller micrometers at several observatories.

The principle of measurement consists of judging alignments and equidistances, and is excellent for the eye, which is especially apt for such comparisons. The results show that an observer adapts quickly, and his internal and external errors are smaller than with filar micrometers; the results are less affected by degradation of the images. This micrometer can be used for pairs fainter than those accessible to interferometers, since all the light reaching the objective is used. As the emergent beams are polarized at right angles, it is sufficient to add a piece of polaroid between the prism and the microscope to make the instrument into a photometer. By turning the polaroid, the intensities of the images can be equalized, two by two. The measurement of the difference in magnitude is reduced to a reading of angles. The observations can be made in special spectral ranges by the use of filters. Muller has made numerous photometric observations with this instrument.

The Lyot-Camichel Double-Image Micrometer

This is a double-image micrometer made with a plane parallel plate of spar inclined to the optical axis of the instrument. The orientation of the plate and its inclination determine the position angle and separation of a couple by comparison with the separation caused by birefringence, but the separation is not proportional to the readings. The designers do not give any other information about this instrument which has been in use at the Pic-du-Midi Observatory (see Camichel 1949). This micrometer, very simple to use, deserves to be better known; it permits very precise measures.

PHOTOGRAPHIC OBSERVATION OF DOUBLE STARS

Only a few words will be said on this subject, which is part of photographic astrometry rather than of visual observation.

Optical Cameras

Photography is not well adapted to the observation of close couples, since the images formed on the plate by the stars are spots larger than the diffraction image. This latter amounts to a score of microns in instruments of large focal length, but the photographic spot often reaches 100 μm. The diffraction image could be magnified before it reaches the plate, but every optical system involves constants that are sometimes difficult to determine; photographic astrometry, therefore, is done at the foci of refractors, whose focal lengths are well known, or at the prime foci of some reflectors.

For separations less than 100 μm, the images impinge upon each other and are more or less confused. In this case, by examining the photographic spot, one can distinguish a maximum of light: the photocenter. The position of the photocenters is neither that of one of the components nor that of the center of mass. Its distance PA from the principal star is

$$PA = \frac{\ell_B}{\ell_A + \ell_B} AB, \tag{3.5}$$

where ℓ_A and ℓ_B are the luminosities of the components A and B, and AB is their distance (fig. 3.12).

The photocenter describes an orbit similar to those of A and B about the center of gravity. A number of double stars have invisible companions that are detectable photographically, thanks to the movement of the photocenter; these are called *astrometric binaries*. The formula (3.5) is not rigorous, since the photographic formation of a double luminous source with unresolved images is not well understood. However, good knowledge of stellar masses would come from the solution of this problem. Some investigators, such as Morel (1969), have contributed recently to its solution. The great specialist in invisible companions is Peter van de Kamp of the United

Figure 3.12 The photocenter. A and B are the centers of the unresolved images of the components. P is the position of the maximum photographic density.

sources angular separation, s, expressed in radians, is given by the formula

$$s = \lambda/(2d)$$

where λ is the wavelength of the light and d is the spacing of the slits, measured between their centers. The yellow-green light to which the eye is most sensitive has a wavelength of about 5500 angstroms, or 0.00055 millimeter. When s is expressed in arcseconds and d in mm, the formula simplifies to

$$s = 56.72/d$$

and is ready to use for measuring close binary stars.

CONSTRUCTION TIPS

I decided to make the rectangular slits 25 mm wide and 90 mm high. Arranged on a screen in front of my telescope's 200-mm corrector plate, the maximum possible distance between slit centers is about 165 mm. This means I can measure angles as small as 0.35 arcsecond. At the other extreme, the closest practical

BRIGHT BINARY STARS

Star	Right Ascension	Declination	Magnitudes		Position Angle	Separation
γ Cen	12h 41.5m	–48° 58'	2.9	2.9	349°	1.12"
ε Equ	20h 59.1m	+4° 18'	6.0	6.3	284°	0.88"
ζ Boo	14h 41.1m	+13° 44'	4.5	4.6	301°	0.85"
η CrB	15h 23.2m	+30° 17'	5.6	5.9	52°	0.85"
R 65	6h 29.8m	–50° 14'	6.0	6.1	268°	0.79"
ζ Cnc AB	8h 12.2m	+17° 39'	5.6	6.0	104°	0.75"
Σ2173	17h 30.4m	–1° 04'	6.0	6.1	323°	0.75"*
γ Lup	15h 35.1m	–41° 10'	3.5	3.6	274°	0.68"
η Oph	17h 10.4m	–15° 43'	3.0	3.5	239°	0.58"
ζ Sgr	19h 02.6m	–29° 53'	3.2	3.4	244°	0.57"
λ Cas	0h 31.8m	+54° 31'	5.5	5.8	190°	0.56"
72 Peg	23h 34.0m	+31° 20'	5.7	5.8	95°	0.53"
α Com	13h 10.0m	+17° 32'	5.0	5.1	12°	0.28"*
φ UMa	9h 52.1m	+54° 04'	5.3	5.4	251°	0.25"
ξ Sco	16h 04.4m	–11° 22'	4.8	5.1	231°	0.20"

These close, bright binary stars are well suited for interferometric or resolution tests with 6- to 12-inch telescopes. Right ascensions and declinations are for equinox 2000.0, but position angles and separations have been calculated for mid-1997 by Roger W. Sinnott from the orbital elements in *Sky Catalogue 2000.0*, Vol. 2. Most separations change slowly, but the two asterisked values actually shrink by 0.1 arcsecond between the beginning and end of the year.

rificed for greater precision. Designed for use with a 12-cm objective; recommended magnification, not less than 25D; recommended test star, mag 1 or 2.

1: disc and rings undifferentiated; image usually about twice the size of the true diffraction pattern; ⎱ Very poor
2: disc and rings undifferentiated; image occasionally twice the size of the true diffraction pattern; ⎰
3: disc and rings undifferentiated; image still enlarged, but brighter at the centre;
4: disc often visible; also occasional short arcs of the rings; ⎱ Poor
5: disc always visible; short arcs of the rings visible for about half the time; ⎰
6: disc always visible, though not sharply defined; short arcs of the rings visible all the time; ⎱ Good
7: disc sometimes sharply defined, and distinct from rings; ⎰
8: disc always sharply defined; inner ring in constant motion; ⎱ Excellent
9: disc always sharply defined; inner ring stationary; ⎰
10: disc always sharply defined; all rings virtually stationary. } Perfect

States. This astronomer has discovered about ten, of which some are probably large planets, a little more massive than Jupiter.

There is, nevertheless, a large class of pairs whose components are sufficiently separated to be easily observable photographically. The advantages of photographic emulsion are that it is more sensitive than the eye and provides a permanent record. In view of the small distance between the components, the plate can be replaced by a film to obtain a large number of negatives. The direction of diurnal motion is determined by photographing star trails. The exposures must be short, several seconds to a minute, to avoid the blurring of light on the film or plate (see fig. 3.13).

A device used by Ejnar Hertzsprung and then by Kaj Aa. Strand consists of a grid of coarse parallel wires placed in front of the objective. The grating of wires gives little spectra on each side of the central image. If a is the distance between the middle of two consecutive wires, the first-order spectra are at an angular distance

$$d = \lambda/a \tag{3.6}$$

from the central image. These small spectra are almost point images, especially if a refractor equipped with a yellow filter is used. For a spacing of 1 cm between the wires, the first-order images are 11".5

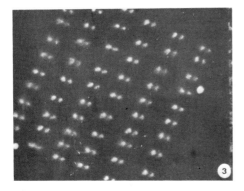

Figure 3.13 Electronographic plate (Ilford G5), showing a series of images of visual double stars. At each side of the plate two small spots can be seen, which serve to fix the zero point of the position-angle scale. Photograph and caption by P. Laques.

from the central image. If the wires are replaced with bars that are bigger and bigger with respect to the space separating them, the brightness of the spectra is increased at the expense of that of the central image. The position of the central image can be determined with great accuracy from those of the first-order images. This arrangement is very useful when the components differ greatly in brightness. The orientation of the grating allows a good determination of the direction of diurnal motion, and the spacing of the wires gives a good value of the distance scale. M. Duruy has successfully used gratings in his observatory at Rouret (Alpes-Maritimes) to measure double stars with a 60-cm reflector.

The Electronic Camera

The electronic camera is based on the photoelectric principle. A particle of light (a photon) is able to tear electrons free from the metal that absorbs it. Thus the photons are changed into electrons. These latter can be multiplied by passing through multistage cells. The resulting beams of electrons are controlled by electron optics and form an image of the object on a photographic plate. This process allows exposures 25 times as fast as those of ordinary photography; it has been used at the Pic-du-Midi Observatory by P. Laques to measure double stars. One obtains the image instantaneously, and thus avoids some of the blurring due to atmospheric turbulence. Finally, the electronic plate has a linear response, unlike classical photography, and this makes good photometric measures possible. All the same, there are difficulties in checking the distortion of the field. Photographs of standard grids allow the determination of constants, in particular the equivalent focal length.

Electronic photography, called *electronography,* has made it possible to obtain pictures of diffraction rings, thanks to short exposure times that are often less than the characteristic duration of random atmospheric turbulence. Besides, this turbulence is "frozen" on the negative: The images of the two components of a double star show the same deformations—which do not prevent a precise measure of their relative position.

MODERN INTERFEROMETRY

The Automatic Interferometer of Wickes

The Fizeau-Michelson principle of interferometry has been taken up again by W. C. Wickes (1970) and then by A. Labeyrie (1973). Wickes conceived an interferometer based on Young's slits. He isolated the image of the star at the focus of the telescope, then made it pass through a rotating, totally reflecting prism that makes the beam parallel. The beam is made to rotate about its axis by the prism, and passes through a screen pierced by two slits. The rotating fringes that result fall on a screen of radial bands, alternately opaque and transparent, which analyze the image by transmitting it to a photoelectric system. In short, the eye is replaced by a faster and more sensitive system. In fact, its performance does not increase the resolving power much, but the system reaches fainter stars than the eye can. Moreover, the information is recorded and can be used at will.

Labeyrie's Speckle Interferometry

In 1970 Labeyrie and his collaborators achieved a major advance by using the full aperture to form the image. First, they studied the formation of the image of a star at the focus of a large reflector, taking into account the atmospheric perturbations. Cinematography allowed them to make a detailed study of the complex structure of this image that looks like a bunch of grapes, in which each grape (or speckle) is violently agitated and has a very short lifetime. They began with the idea that the image is the effect of interference between the incident beam and random spatial fluctuations of phase. A speckle is nothing but the Fourier transform of the telescope pupil; it is the diffraction image of the star. For a binary, therefore, each speckle contains the information that the whole "bunch" is double, but the displacement is very small—smaller than the diffraction image. Interference fringes appear if diffusion of a coherent beam is observed through a negative of a "bunch" (a double-star image). It can be shown that the separation is

$$\rho = \lambda_0 (f'/s) f, \tag{3.7}$$

where λ_0 is the wavelength of the coherent beam, s is the space between the fringes, and f' and f are the equivalent focal lengths of the telescope and of the Fourier-transform device.

The technique has been used by its inventors at the 5-m Palomar reflector. They took more than 100 photographs of the star to be studied, with exposures of 0.0001 second. The equivalent focal length is 889 m, which gives a scale of 2"5 per centimeter. Each photograph is viewed in coherent light, and identical fringes of low contrast are obtained (fig. 3.14). Adding all these photographs on a single plate augments the contrast. The same observations are made for a reference single star. Each point of the intensity profile of the composite image of the double star is compared with the similar profile of the reference star. The result is a visibility curve for the fringes analogous to that in the theory of Fizeau and Michelson. Autocorrelation methods enable the image of the couple, including the brightness ratio, to be reconstituted.

This procedure is long, but it returns the same information that one could obtain from perfect images. It enables the resolution of the 5-m reflector to be reached (0"02) for differences of brightness up to five magnitudes (1972). Labeyrie and his collaborators have revealed the companions of several spectroscopic and astrometric

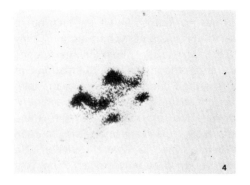

Figure 3.14 Magnified image of a double star of small Δm and separation about 0"9. Two well-resolved granules are seen, which allow easy astrometric measurement. Photograph and caption by P. Laques.

69 Measuring Instruments

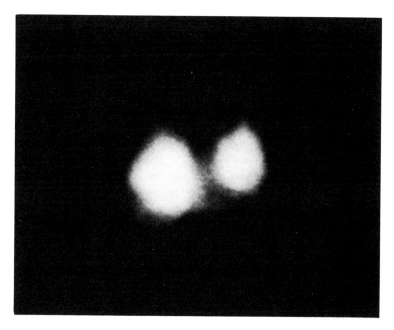

Figure 3.15 Photograph of a double star, ADS 2709 = 0Σ65 $3^h47^m3 + 25°26'$ (1950). One-second exposure through the Nice 50-cm refractor, 1967.818, 203°9, 0".65, 6^m5-6^m8. Focus 22.5 m, magnification 300×. Separation of the centers on the plate 72 μm. Nice Observatory photograph by E. Fossat.

binaries, such as π Persei, γ Persei, δ Scorpii, χ Draconis, and some new double stars such as β Cephei (fig. 3.15).

Success in observing double stars with the full resolving power of the large reflectors has thus become a reality. The program committees of the large telescopes have understood the importance of this work; American astronomers have decided to use regularly one of the transatlantic monsters, the 4-m Mayall reflector at Kitt Peak National Observatory. There, Lynds and McAlister observe double stars of all kinds regularly in this way, and have perfected the details of Labeyrie's technique. Many spectroscopic couples have been resolved, such as 12 Persei, η Orionis, and η and θ Virginis.

The performance can be improved by using two telescopes separated by some tens of meters, or a group of telescopes in a regular polygon, which act, for the purpose of resolution, like a single instrument of the same size as the group. An installation of two coupled telescopes separated by about 10 m is under construction at the Centre d'Etudes et de Recherches Géodynamiques et Astronométriques near Grasse.

Hanbury Brown's Intensity Interferometer

R. Hanbury Brown, J. Davis, and L. R. Allen obtained, by a slightly different technique, no less spectacular results at Narrabri Observatory, in Australia, in 1971. They used two movable mirrors on a circular railway track, and channeled the luminous flux, changed into electric energy, towards a central correlator. Unfortunately, the technique is limited to very bright stars. This instrument has been able to resolve α Virginis, and to determine its distance from the sun (84 parsecs).

OBSERVATION OF DOUBLE STARS BY OCCULTATION

Everyone knows that a star occulted by the limb of the moon disappears suddenly because of its very small angular diameter. In general, the two components of a binary are occulted one after another in a way determined by their separation and their position relative to the lunar limb. This phenomenon is not instantaneous, although it can scarcely be made to yield information visually.

The speed of response of modern photoelectric photometers is such that it is now possible to follow in detail the phases of an occultation, especially those that are grazing. If a double star is occulted, the analyzer will show two diffraction traces, more or less confused but separable (figs. 3.16, 3.17). From these traces, the brightness of each component and their separation projected on the lunar diameter can be deduced. If the occultation is observed from two different places, then the separation and position angle can be determined with an accuracy of 0.001 arc-second. It is useful, therefore, to observe occultations from a number of observatories at widely separated locations. An association has been formed with this aim, and it publishes a journal, *Occultation Newsletter*. D. Evans, D. Dunham, J. Africano, and others at the McDonald Observatory in the United States have found scores of very close binaries. One of

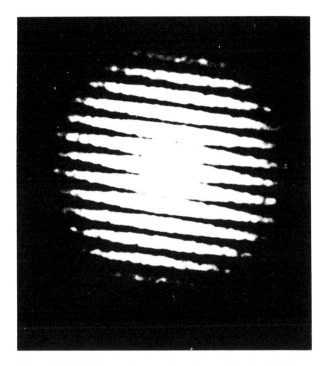

Figure 3.16 Interference fringes for κ Ursae Majoris. The fringes are obtained by combining 50 plates taken with the 4-m telescope of Kitt Peak National Observatory. The separation is 0."272. Photograph by H. A. McAlister from *Sky and Telescope*, by permission of the author and the editor.

the most remarkable is Atlas, in the Pleiades, which has a fifth-magnitude companion separated by 0."004. It should be understood that occultations, like interferometry, permit the measurement of the apparent diameters of stars, of stallites in the solar system, and of minor planets.

OBSERVATION OF DOUBLE STARS BY PHOTOELECTRIC SCANNING OF THE IMAGE

This technique, related to that of occultations, was developed and used by Karl D. Rakos in Vienna and Otto G. Franz at Flagstaff. The

Figure 3.17 Image of β Cr B obtained by autocorrelation. North is towards the bottom. The companion can be seen at a position angle of 150° at 0."30 from the primary. Plate by A. Labeyrie, Hale Observatories.

method consists of placing in the focal plane a screen pierced by two narrow, mutually perpendicular slits (fig. 3.18). The screen can be turned to orient the slits, and it can also be moved in its own plane along one of the bisectors of the angle between the slits. A plane parallel plate placed in front of the screen allows the observer to look at the star and to guide it on the slits.

To observe a double star, the slits are oriented in such a way that one of them contains both star images; then the screen is moved. The two components first move simultaneously through one slit, then successively through the other. The light profile is recorded in the form of an intensity curve. The profile corresponding to the first slit is that of a single star, and serves for comparison. The task is to recover, from the record, the profile of each component and their relative intensities. Either Fourier transforms can be used, or parameters can be calculated for two functions whose sum best represents the profile of the double star. The orientation of the slits already provides an approximate value for the position angle. If the components are too close to be separated, several records are made; the result is obtained after a rather long analysis. The advantage of

73 Measuring Instruments

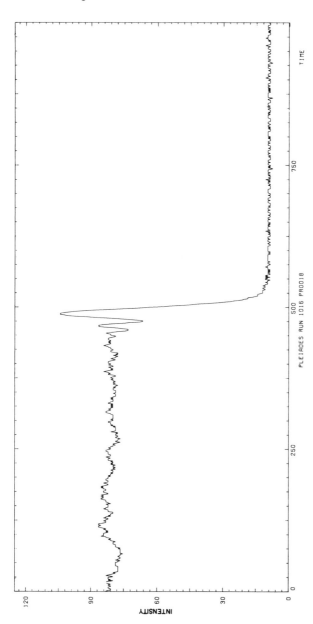

Figure 3.18 Light curve of an occultation of Alcyone, September 11, 1973. The profile is that of a single star. The abscissa gives the time in milliseconds; the intensity scale on the ordinate is arbitrary. Diagram by D. S. Evans, University of Texas.

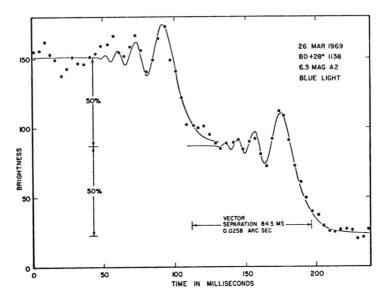

Figure 3.19 Light curve of lunar occultation of the star BD +28° 1138 (6.3) Note the two sets of fringes. The two components are separated by 0.″0258, or 84.5 milliseconds in time. Their magnitudes are 6.ᵐ7 and 7.ᵐ5. Diagram by R. E. Nather and D. S. Evans, *Astrophysics and Space Science.*

this technique is that it gives at the same time the coordinates of the couple and a good measure (to some thousandths of a magnitude) of their difference in brightness.

BIBLIOGRAPHY

Arend, S. Réglage pratique de l'équatoriel visuel et de l'astrographe. Observatoire Royal de Belgique monograph no. 2, 1951.

Baize, P. "Valeur comparée des mesures filaires et non-filaires dans la mesure des distances d'étoiles doubles." *Journal des Observateurs* 32 (1949): 97.

Brown, R. H., J. Davis, and L. R. Allen. "The Stellar Interferometer at Narrabri Observatory I." *Monthly Notices of the Royal Astronomical Society* 137 (1967): 375.

Camichel, H. "Mesures d'étoiles doubles faites au Pic du Midi." *Journal des Observateurs* 32 (1949): 94.

75 Measuring Instruments

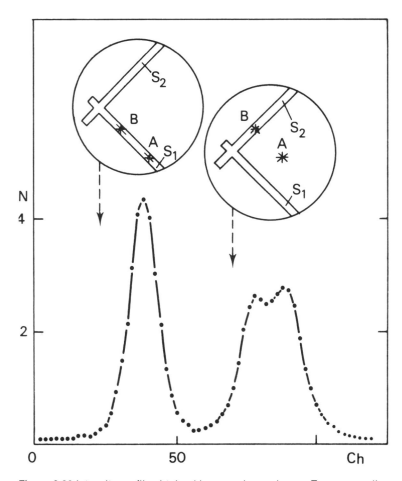

Figure 3.20 Intensity profile obtained by scanning an image. Two perpendicular slits S_1 and S_2 are cut in a screen placed in the focal plane. The two images of the components traverse one of the slits at the same time, giving maximum intensity. They traverse the other slit one after the other. The intensity profile and the orientation of the slits allow the quantities required to be found. From Rakos 1974.

Danjon, A. "Description et théorie d'un micromètre interférentiel demi-onde." *Annales de l'Observatoire de Strasbourg* 3 (1937): 181.

Evans, D. H., R. H. Brown, J. Davis, and L. R. Allen. "A Study of α Virginis with an Intensity Interferometer." *Monthly Notices of the Royal Astronomical Society* 151 (1971): 161.

Franz, O. G. "A Photoelectric Area Scanner for Astrometry and Photography of Visual Double Stars." *Lowell Observatory Bulletin* no. 154 (1970).

Jonckheere, R. "Un micromètre à fils obliques." *Journal des Observateurs* 33 (1950): 57.

Labeyrie, A. "Observations interférométriques au Mont Palomar." *Nouvelle Revue d'Optique* 6 (1975): 15.

McAlister, H. A. "Speckle Interferometric Measurements of Binary Stars." *Astrophysical Journal* 215 (1977): 159.

Morel, P. J. Contribution à la détermination photographique du rapport des masses d'une binaire visuelle. Thèse de spécialité, Nice Observatory, 1969.

Muller, P. "Sur un micromètre à double image, ses possibilités et quelques questions connexes." *Bulletin Astronomique* 14 (1947): 13.

Rakos, K. D. Photometric and Astrometric Observations of Close Visual Binaries. Colloquium for the Inauguration of the Astrometric Telescope of Turin Observatory, 1974.

van de Kamp, P. *Principles of Astronomy*. San Francisco: Freeman, 1967.

Wickes, W. C., and R. H. Dicke. "An Automatic Interferometer for Double Star Observations." *Astronomical Journal* 78 (1973): 757.

4
SOME PRACTICAL ADVICE

I have reviewed some of the observing methods, and stressed particularly the filar micrometer, with which the greatest names in double-star astronomy have worked. Most of the great observers never used anything else. This chapter will describe in detail what a visual observer should do, both when preparing for and when spending a night in the dome. I will set out the fruits of my own experience of more than a quarter of a century's visual observation of double stars. My aim is to guide the young astronomer who is beginning to observe. These things are not learned like theories in physics, but are acquired over the years, like the trade of a craftsman. Everything that an astronomer does in the dome carries his signature; he works in order to publish his observations, which will be judged by his peers, sometimes long afterward. He knows that all our astronomical knowledge, even the most abstract theory, is based on observation. He knows also that a good measurement will be used one day and an erroneous observation will always be unmasked.

PREPARATION FOR AN OBSERVING NIGHT

Observing nights would be quiet sessions of laboratory measurement if the atmosphere did not severely limit both the quality and quantity of the performance to be expected from a night. The perturbations created by the atmosphere are called *seeing*. This is the principal obstacle to observation, but it is enough. Seeing, which we will study in some detail in the following pages, acts at two levels: at high altitudes as a result of strong winds in the upper levels of the atmosphere, and near the ground as a result of convection currents caused by the exchange of warm and cold air. The astronomer cannot do anything to diminish the high-altitude seeing, but he can act at ground level, particularly by creating air currents in the dome.

It is possible to know in advance, during the day, by observing natural phenomena, whether or not the night will be good. Total absence of wind is a good sign. If the leaves stay still at the ends of branches, even eucalyptus leaves that flutter in the slightest breeze, one can hope for good images. At Nice, the sea provides an excellent criterion: It should show large flat surfaces resembling those of oil, elongated like rivers; no white breakers on the shore; and a barely visible horizon. The sky ought not to be too blue at the zenith. If the smell of vegetation stays near the ground, and the smoke from house chimneys rises only slightly, or even falls back again toward the ground, the chances are that the images will be good. Of course, one can look at a star through the telescope in full daylight to test the quality of the sky, although daytime seeing differs from that at night. It is known that, in principle, the planets do not twinkle; but if the seeing is bad, Venus, Jupiter, Mars, and Saturn twinkle very slightly. Their twinkling is particularly noticeable to a short-sighted person who takes off his glasses. If he sees no twinkling when he has done that, that is a certain sign of good seeing.

The slit of the dome should be opened an hour before observing begins, and the instrument put in a nearly vertical position with the valve open. Long refractors are like balloons: They trap warm air, which rises and stays behind the objective. The air is cooled again by the tube and the objective, and forms slow eddies. This effect is particularly marked in the 18-m-long great refractor at Nice. It is one of the reasons why the tubes of refractors should be kept horizontal during the day. For the same reason, one should avoid letting the slits and sides of domes receive heat from the setting sun. It is good to ventilate the room before observing, in order to expel all the warm air. Avoid having too many people under the dome before observing; their body heat will not create a favorable environment for good seeing.

Danjon advised sprinkling domes in order to cool them; this is not always possible. Plenty of grass kept well watered around a dome, and copses of shrubs, moderate the heat exchange. There should be no parking lots near the dome opening, as cars are sources of heat. In general, the natural vegetation of the site should be protected. Deforestation and too many buildings favor the creation of ascending currents of hot air at the beginning of the night. On mountain summits and high plateaus the image quality is better than at the sides. Double-star observation requires a still atmosphere; a light haze is not a nuisance. Strong winds, which clear the air and give a deep blue sky, are sources of bad seeing.

PRECAUTIONS TO TAKE WITH THE TELESCOPE

There are a number of instruments in the dome, essential for good observations, whose quality depends on the care with which they are maintained. In the first place, the objective should be clean; otherwise the starlight is absorbed and diffused. Do not hesitate, therefore, to clean the side of the objective facing the sky (crown) with alcohol, and even the inner face (flint) if it is accessible. A very clean worn-out fine linen rag is perfectly suitable. Rub until all marks have disappeared, applying alcohol several times, as necessary. Soot from heavy oils is a great enemy of glass, because it sticks and can penetrate old flint glass. The objective should be inspected at least once every two months. Similar precautions should be taken to keep the eyepieces clean, but they are often coated to reduce reflections, and therefore cannot be wiped but should be gently brushed with a camel-hair brush.

The inside of the tube of a refractor ought not to reflect or diffuse any light towards the objective. Check that the inside of the tube is in good order by shining the light of the sky on it. Sometimes the inside of the tube has long grooved pipes, with a space between them and the tube, to prevent light from the lamps placed inside the telescope or near the objective from being diffused. The pipes, which make the instrument heavy, can be removed, leaving only two diaphragms, provided that sources of light inside the instrument are avoided as far as possible.

Many of the instruments used for measuring double stars are fairly old refractors, still on German mountings designed to permit observation of all parts of the sky. These instruments, whose principal parts overhang, must be perfectly balanced to avoid the risk of

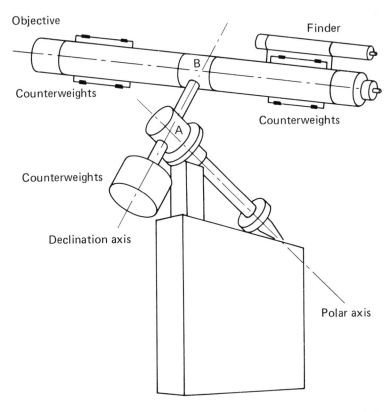

Figure 4.1 Balance of an astronomical refractor. **A:** center of gravity of the moving parts. **B:** center of gravity of the telescope.

serious accidents. The balancing is done in two stages. First, check that the center of gravity of the tube is at the same height as the declination axis. This requires a careful adjustment of the counterweights (fig. 4.1); the check is made for several positions of the instrument, with a spring dynamometer. Next, check that the center of gravity of the moving parts is at the intersection of the polar axis and the declination axis. As an example, the moving parts of the large refractor at Nice weigh 7 tonnes and are balanced to within 2–3 kilograms; pneumatic systems clamp the instrument and ensure complete safety. Old instruments also have counterweights to relieve the roller bearings. The instrument must be balanced anew each time auxiliary apparatus is removed. Forgetting this elementary rule has caused catastrophes that have cost human lives.

Mountings made at the end of the last century were often provided with graduated silver circles. The finest divisions were each arcminute in declination and each 20 seconds in time. These circles should be cleaned as frequently as the objective, but no deposit of polish should be left on them. Preferably, they should be read with the help of an index made of nylon thread; avoid pointers which cast a noticeable shadow. The circles are read through small astronomical telescopes and totally reflecting prisms. Keep these old arrangements, modifying only the illumination; they enable the telescope to be set on objects to within a fraction of a minute of declination, and a few seconds in right ascension, which is necessary. In fact, the double-star observer has available a stellar field of 5–6 arc-minutes. Faint stars are invisible in the finder; he can count only on his instrument to identify them. Even though there are electronic setting systems, the observer will do well to keep the old circles, which have the advantage of reliability under every test.

ATMOSPHERIC SEEING

Before we deal with the practical problems of observation, it is convenient to study the principal cause of the destruction of the image: atmospheric seeing.

Light rays are subject to six main perturbations by the atmosphere, which refracts, absorbs, diffuses, disperses, breaks up, and smears out the light. Each of these contributes to the deterioration of the images and often makes them unrecognizable and useless. In general, refraction, absorption (extinction), and diffusion are of less importance for high-resolution observations. Dispersion is nothing but differential refraction: It makes the stars look like little spectra. In practice, this phenomenon is noticeable only a long way from the zenith, and it is weakened in refractors. It can be reduced by interposing a yellow filter. By contrast, image motion and blurring are responsible, in most cases, for image degradation. Images move because plane wavefronts of random inclinations, caused by slow variations in the refractive index of the atmosphere, pass in front of the objective. Blurring is produced by a deformed wavefront which can no longer be considered as plane. Image motion is not seen through large apertures, because the wave, deformed by rapid variations in the refractive index of air, is never plane over the whole objective. With a small aperture, well-formed images can be seen on some nights; but they move about a mean position, whereas in a big instrument they appear stable but blurred (fig. 4.2).

It is the combination of image motion and blurring that is called seeing. Mathematically, it is defined as the mean angular deviation between the light ray and its theoretical direction. The deviation varies in a random manner inside a small cone of very small semivertical angle (at most a few arc-seconds). Danjon defined a scale of seeing giving the order of magnitude of this angle as a function of the appearance of the image. When the seeing reaches the angular value of the resolving power, the diffraction image is much altered; its form tends to disappear, and the rings vanish. When the

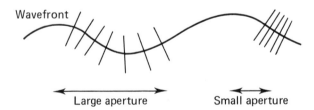

Figure 4.2 Atmospheric seeing. A wavefront impinging on a small objective can be considered as plane, but not one impinging on a large objective. In the first case the images move but are distinct; in the second case they are blurred but stable.

seeing is considerably less than the resolving power, for example about half, the Airy disk remains easily visible; the instrument can perform to its full capacity. For example, an instrument of 50 cm aperture, with a resolving power of 0.24 arc-seconds, cannot tolerate seeing above 0.12 arc-seconds without the images it forms being noticeably affected. Since the seeing is rarely less than 0.1 arc-seconds, it is obvious that large apertures are seldom able to work at full capacity.

It is important to notice the quality of the images, both when observations are begun and while they are in progress. In general, if useful observations are to be made, the diffraction rings and the central disk should be visible. Correct observations cannot be made when seeing enlarges the disk to the point that it is unrecognizable. Vigorous motion of a fairly pure image is far preferable to blurring, because the eye and the brain filter out the motion and keep the information, provided that it appears from time to time during the observation. Many observers have acquired the habit of noting the image quality on a scale from 0 to 5, going from very blurred images to pure diffraction patterns. The quality varies from one minute to the next, even in the course of a very good night. It is at sunset, at the time that the wind regime changes toward the night breeze, that the best images are seen.

Observation of *shadow bands* gives valuable information about the direction and speed of the winds and the thermal eddies inside the instrument. To see these, set on a bright star, remove the eyepiece, and put the eye near the focal plane. The objective is seen illuminated and traversed rapidly by striae, or shadow bands, which resemble a little those seen at the bottom of a swimming pool when the surface is rough. These billows are caused by sheets of air of variable refractive index (fig. 4.3). Next, insert the eyepiece and pull it out some millimeters, so that the image of the star is out of focus. The instrument is focused instead on a plane some kilometers away, whose appearance resembles that of a flag fluttering in the wind. By this procedure the altitude of the winds can be estimated. If they are very high, there is little hope of improvement in the course of the night.

Seeing of instrumental origin must be distinguished from that of atmospheric origin. The first causes low-frequency eddies, which

84 Practical Advice

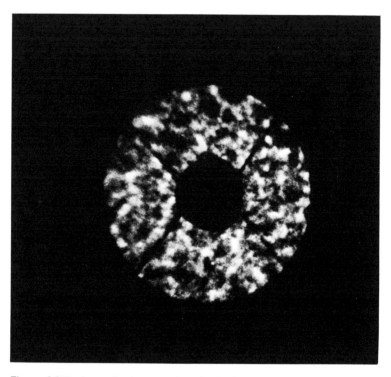

Figure 4.3 Photograph of shadow bands obtained with the 152-cm reflector of the Observatory of Haute-Provence. The exposure time was 1/1,000 second, with the brightness amplified 80,000×, on Tri-X film. If there were no turbulence, a uniformly clear patch of light would be seen. Note the silhouette of the Cassegrain secondary mirror and its support. Photograph by F. Martin and J. Borgnino, University of Nice.

can be modified by opening the windows; the second causes uniform blurring of the image, which is larger the farther away the origin of the perturbation.

As the appearance of the image deviates from the diffraction pattern, difficulties are experienced in the measuring. Bisection by the micrometer thread is badly judged, as also is parallelism. Double-image micrometers and interferometers are less sensitive to degradation of images. Perfect nights are very rare; astronomers are constantly constrained by the seeing, all the effects of which (particularly image motion and blurring) add up. The result is that nights are seldom of a quality favorable for double-star observation. The astronomer must be on the lookout for good times. The hours spent in the dome are very often hours of waiting. The astronomer who tries to profit at all costs from mediocre conditions risks making unusable observations and, if he is searching for new pairs, not discovering the most difficult and therefore the most interesting binaries.

PRACTICALITIES OF VISUAL OBSERVATION OF DOUBLE STARS

The first thing to do when you arrive in the dome to observe is to check the reading of the astronomical clock. The sidereal time is given nowadays by electronic computers, which are not to be trusted because they are fragile, can break down, and sometimes indicate a fantastic time. Keep carefully the old weight-driven clocks, which never get out of order, even after the worst thunderstorms. It is a good idea to set the weight-driven clock to a time slightly different from the sidereal time, according to the date of the equinox chosen for the coordinates of the stars. For example, if the coordinates are given for 1950, set the clock at the time for the equinox of 1950, and the correction to the clock is almost zero throughout the night.

All lights in the dome should be put out during observation. Remember that the eye must stay in the dark for 40 minutes before reaching its full potential. The stars to be observed should be written out on a list, or in a notebook, which can be illuminated, if necessary, by a small neon night light. The astronomer who walks around at night to do his work should be content with the light that falls from the sky. It is a good way to accustom his eye to the work that is waiting for him.

Observing stars is not dangerous to the eye, but it is well to take precautions for bright objects. In effect, all the luminous energy collected by the objective passes through the exit pupil, which is smaller for higher magnifications. As a result, the cornea can receive, on a small part of its surface, the entire light flux passing through the objective. This flux has the value $C(M/M_n)^2$ times that received by the naked eye. The illumination received on a point of the cornea is increased in this ratio, which can reach 5 million for large instruments. Not all the flux reaches the retina; besides, the part that does is distributed over its surface. When looking at the moon or a planet one will not have the impression of being dazzled, but the cornea is subjected to luminous bombardment that can eventually tire the eye. Finally, the very thin pencil of light that enters the eye is affected by faults within the eyepiece, but this does not bother a seasoned observer.

Double-star observation requires a long and tedious training; that is the chief obstacle to recruiting the young. The fundamental difference between photographic and visual observation is that in the first case it is the instrument that does the work, but in the second it is the astronomer. The latter improves himself by practice, but he needs eight or nine years' experience before he becomes a recognized observer. If he changes his objective, he will require a year to accustom himself to the new one. This is the time that the author, quite used to observing with the Nice 50-cm refractor, required when it was replaced with one of 38 cm. He needed more time with the large 76-cm refractor. Each objective has its own personality; none is identical with another. In fact, the secondary spectrum, the focal length, the achromatism, and the tempering are never the same. Neither is one's reaction to the atmospheric seeing of the site, the structure of the dome, or the objective. The first observations with a new refractor are always deceiving, even if one has already broken with routine, because the eye must accustom itself to the conditions of each instrument. These reasons explain the small number of regular observers, whose perseverance has been put to the test. André Danjon, speaking of double-star observation, said it was necessary to have it in the blood.

Without exaggerating their importance, certain natural dispositions are needed if you are to become a good observer. You should work physically relaxed, without effort or muscular tension. You do not need to close an eye while observing, since the one you are not using is in front of the micrometer casing, which acts as a screen. Your head, thrown back, can rest effortlessly on your shoulders; a long neck is a disadvantage. You should be upright while observing, bending your legs to bring your head to the right height. A sitting position is not advised because it loses time. Sophisticated observing chairs, such as are sometimes to be found in domes, should be put in the museum. A good observer should feel at ease in all positions, a trait that can be acquired only very slowly. Beginners experience real difficulty with these exercises. An assistant's help is unnecessary if the astronomer knows how to organize himself well; he can dictate observations onto magnetic tapes or, as at Washington and Nice, record them automatically. That way, he avoids the frustration of a secretary who goes to sleep without warning.

Observations south of the zenith should be made with the tube west of the pier, in such a way as to see objects before their meridian passage—which is preferable for the quality of the images. If circumpolar stars are observed, it is convenient to do so with the tube to the east; then the astronomer and the instrument have the same relative position. Old telescopes are designed for observing with the tube to the east, and several astronomers have kept that habit, but in this position the stars are observable only after they have crossed the meridian.

When observing, you should act quickly but not brusquely. The star image should not be moved (otherwise, physiological effects will make it look deformed); it ought to be close to the middle of the field. The eye, always the same, should scrutinize the image rapidly, accustoming itself to the appearance of single stars as the light rays fluctuate. Do not hesitate with the bisections or the estimates of parallelism, which should always be made with the line of the eyes parallel or perpendicular to the direction of the components. It is an advantage to make five settings of the position angle and four of the double distance. There is always an optimum magnification to use, which is determined by the often variable conditions and the characteristics of the object being observed. Therefore, you should change the eyepiece very often, a little as the driver of a car changes gear according to his route. It is even often advantageous to measure the position angle and the separation with different magnifications. To bisect the disk one has to be able to see it; this requires a high magnification, whereas parallelism between the line of components and the thread is best judged with sharper images. An observation of a double star should take no longer than 5 minutes. Taking account of time for adjustments, a rhythm of eight measures an hour is excellent with a properly mounted instrument.

The more the aperture is increased, the more sensitive it becomes to seeing. The writer's experience with refractors of apertures from 38 to 102 cm enables him to make some comparisons. If the images are perfect through an aperture of 38 cm, they are nearly perfect through one of 50 cm and very good through one of 76 cm. As mentioned, the atmosphere limits the magnification to a particular value, whatever the aperture. For example, if conditions do not permit exceeding a power of 900× with a 38-cm refractor, you cannot exceed it with one of 76 cm, but you can go up to that value. In such conditions, the large aperture does not enable you to observe pairs too close for the small one. If conditions are still worse, experience shows that the decrease in the effectiveness of a large aperture becomes very important. Double stars with a large difference of brightness between the components will no longer be observable, even though a smaller aperture will still show them.

At the limits of the instrument, the eye experiences difficulties, not only in measuring, but also in judging. For example, faced with a difficult pair, one becomes incapable of deciding whether the components are of equal brightness and very close, or of very different brightness and farther apart. Some position angles are more difficult to measure than others, particularly those that require a more uncomfortable position of the head, around 130° or 310°.

Observations dictated on magnetic tapes, or printed in a roll, can be reduced the next morning and transcribed into a file. Do not hesitate to describe your observations and to add redundant comments noting the degree of difficulty of the measurement, or astonishment at an unexpected sight.

This section ends with a word of warning to those who are getting ready to visit an astronomer in his dome: Do not forget that an astronomer who observes perfect images visually is a wild beast who devours his prey. Do not disturb him under any pretext. Let nature take its course!

OBSERVING ARTIFICIAL CELESTIAL OBJECTS

At the Palace of Discovery (in Paris) and in some astronomical laboratories, optical montages have been constructed that reproduce the appearance of some double or multiple stars, planets, nebulae, and star clusters. These montages can be observed, under conditions quite like the real ones, through imitation telescopes. The apparent motions of the stars and planets can be shown by mechanized celestial vaults or planetaria. These montages make it possible to show the sky to visitors at any time. Naturally, nothing is as good as the examination of a real celestial object, because the atmospheric seeing cannot be reproduced, nor the innumerable variety of celestial bodies with their colors; but you can have a good illusion of observation in this way, familiarize yourself with the technique, and try a few measures.

The amateur who has a refractor or a small reflector will find it useful to make some artificial double stars for himself, to check the quality of his instrument. Nothing is easier than to make an artificial double star; the illusion is perfect. It is sufficient to look at the image of a small lamp, or of the sun, reflected by a polished steel ball bearing situated far enough away for the apparent diameter of the image to be negligible. With a 10-cm objective, a ball bearing 4 mm in diameter could be used; place it about 100 m away and look, in bright daylight, at the reflected image of the sun. Better still, at nighttime, use a small pocket lamp a few decimeters in front of the ball bearing. You will have a stable stellar image, unaffected by seeing, that can easily be examined comfortably, without twisting your neck. Reflections from two lamps, side by side, will give a beautiful double star. The separation can be varied at will, up to the limit of resolution, and even differences in brightness can be created by moving one lamp with respect to the other.

I have often used this method to study the quality of small telescopes acquired by amateurs. Let R denote the radius of the ball bearing, L the distance between the lamps, h their distance from the ball bearing, and D that of the objective from the ball bearing. The separation of the artificial pair in arc-seconds can then be written as

$$s = \frac{RL}{hD} \times 10^5.$$

With ball bearings of 4 mm diameter, and lamps separated from each other by 10 cm, 1 m away from the bearing, which is itself observed from a distance of 100 m, the separation is 0.2 arc-seconds. This arrangement will enable you to mimic a number of known pairs, and to instruct your friends at observing sessions, even in cloudy weather. Moreover, in the town or in the country, artificial double stars abound in full daylight. Every rough surface reflecting sunlight is an artificial star. Insulators on power poles or telephone poles, window panes, ceramic tiles, and mica particles on granite walls all provide lots of stellar images. The old ruined windmills that stand out on the horizons of the French countryside make excellent artificial planets. If we have the fortune to have a distant horizon, we can find one, 20 or 30 kilometers away, whose apparent size corresponds to those of the large planets. In the sunlight, the circular walls of these buildings show a terminator, and the spots and cracks of their ruins make excellent planetary details. Their advantage is that we can go to the place and photograph the object under the same conditions of lighting in which we observed it. We can compare photographs, or drawings, made through the telescope with what we see at the site, thus simulating the space probes that show us the places we have photographed with our earthbound telescopes. It is a good way to familiarize yourself with a telescope; I have often used it at Puybeillard, in the Vendée, my small holiday observatory, where I own part of a wood. The countryfolk are puzzled, and look at me with distrust, when I arrive on some height to examine one of their old windmills, and show an unusual interest in any spots or rough patches that look like high mountains or (lunar) seas.

BIBLIOGRAPHY

Barocas, V. "Atmospheric Seeing." *J. British Astronomical Association* 82 (1972): 279.

Couteau, P. "L'Osservazione Astronomica ad alta risoluzione." *Pul. Instituto Nazionale Di Ottica* 4 (1972): 883.

Kozhevnikov, V. I. "Durnal Visibility Conditions Near the Earth's Surface." *Solnech Dannye Bjull. S.S.S.R.* no. 12 (1971): 104.

Martin, F., J. Borgnino, and F. Roddier, "Localisation des couches turbulentes atmosphériques par traitement optique de clichés d'ombres volantes." *Nouveau Revue d'Optique* 6 (1975): 15.

Rosch, J. (editor). International Astronomical Union Symposium No. 19: "Le Choix des Sites d'Observatoires astronomiques." *Bulletin Astronomique* 24, fasc. 2 (1963): 85.

Roddier, C. Etude des effets de la turbulence atmosphérique sur la formation des images astronomiques. Thesis, University of Nice, 1976.

Vernin, J., and F. Roddier. "Detection au sol de la turbulence atmosphérique par intercorrélation spatio-angulaire de la scintillation stellaire." *Comptes Rendus Acad. Sci. Paris* (1975).

Westheimer, S. "Image Quality in the Human Eye." *Optica Acta* 17 (1970): 641.

5
THE IDENTIFICATION OF STARS

DIFFICULTIES IN IDENTIFICATION

It is of fundamental importance to be able to determine properly which objects you have observed, and to know how to find them in the catalogues. Coordinates of double stars are usually given to the nearest arc-minute in declination, and the nearest 0.1 minute in right ascension. This is generally sufficient, if the observer can trust the coordinates furnished him for his program. To be able to do that, he must assume that the discoverer has done the work of identification well. Sometimes the pair sought has not been observed for decades, perhaps not even since its discovery. Not seldom, then, the desired object cannot be found. More often, slight errors in the coordinates can result in stars being confused, and can end in an inextricable mixture of useless observations. The double-star observer is too often passive. He ought to verify the identity of stars each time he has any doubt; that is to say, whenever he has difficulties finding the star at the position indicated, or even when he sees several pairs in the field of his instrument or in its neighborhood.

Every investigator whose work takes him often to the dome, or even involves him in listing objects in a file, ought to know the basic concepts of star catalogues. The brightest stars (about 3,000) are listed in many catalogues (there are at least seven principal ones), in each of which they have different denominations, not counting their names or constellation letters. More than 200,000 stars are presented in five catalogues. Each one, therefore, has several labels, which must not be confused. Every astronomer ought to be able to obtain the coordinates from the labels, and vice versa. As we shall see, the task becomes more and more difficult as we descend the scale of brightness. The double-star observer is more concerned by these problems of identification, because he is obliged to set on his star with high magnifications; he has only a small field, usually containing the star he seeks all alone, lost in the night. If the appearance of the double star conforms to what he expects, he is content to measure it; but if the star is single, or has become unrecognizable because of its orbital motion, he must make sure of the correct identification of the object. This becomes of prime importance when he sets out to discover new pairs; he must know in advance to which stars he should give special attention, and be sure enough of himself to correct any errors he may find in the catalogue.

The history of catalogues is long, and is not a subject for this book. Each generation since Kepler has struggled to know the positions of the stars better, and has produced more and more precise catalogues. These labors are often only of historic interest. Catalogues can be classified into four principal groups: fundamental catalogues, intermediate catalogues, general catalogues, and special catalogues.

FUNDAMENTAL CATALOGUES

The purpose of these catalogues is to give the absolute positions of a certain number of stars, called fundamental stars, judiciously distributed over the celestial sphere. They serve as reference stars to determine the fundamental planes (equator and ecliptic). In their turn, the fundamental planes serve to determine the positions of other stars.

The first fundamental catalogue was compiled by Auwers in 1879 and published by the German astronomical society (Astronomische Gesellschaft, or AG). It is designated by the initials FC, and contains 539 northern stars. The second fundamental catalogue (NFK) and the third (FK3) were published by J. Peters in 1907 and A. Kopff in 1938, respectively. The number of stars was increased to 1,535, including all those in Auwers's catalogue. Finally, in 1963, the fourth fundamental catalogue (FK4) appeared. It is a revision, by W. Fricks and A. Kopff, of the FK3, and gives positions to $0\overset{s}{.}001$ and $0\overset{''}{.}01$.

INTERMEDIATE CATALOGUES

Like the fundamental catalogues, intermediate ones contain a certain number of stars—but a larger number. The precision of the positions is less, but in the most recent catalogues there are up to 300,000 stars. Until 1920, intermediate catalogues were made with the help of meridian observations. It is only since 1928 that astrographic cameras have been employed systematically.

The oldest catalogues that have contributed to the calculation of the present positions of the stars are those of Halley, Flamsteed, Lacaille, and Bradley, which appeared in 1679, 1691, 1763, and 1817, respectively. Many double stars in Burnham's catalogue are identified by reference to these old compilations.

Special mention should be made of J. de Lalande's catalogue, still used because it is always useful for referring to stars of large parallax or of nonlinear trajectory. The observations were made from 1789 to 1799, with a small refractor of 7 cm aperture, at the observatory of the Ecole Militaire in Paris. Even the turmoil of the Revolution did not prevent de Lalande from observing the positions of 47,390 stars in 11 years. Paradoxically, it was not a French observatory, but a British scientific foundation, that, in 1847, reduced and published the observations of this astronomer. In this work, positions are given for the equinox 1800 with precisions of $0^s.01$ and $0''.1$. Lalande's catalogue was very much in advance of its time, both in the number of stars it contained and in its precision.

Other catalogues, notable in their time, appeared in the first half of the nineteenth century; in particular those of Piazzi and Bessel. Thanks to these catalogues, the proper motions of numerous stars have been determined since the end of the last century. Groombridge's important catalogue, published in 1838 and still used these days, should also be mentioned. This is a catalogue of 4,243 circumpolar stars, made in London at a private observatory.

After 1865, astronomers came to feel the need to know precisely the positions of hundreds of thousands of stars, particularly in order to help in the discovery of planets and comets, and to enable their trajectories to be easily determined. The Astronomische Gesellschaft organized a chain of twenty observatories across Germany, Russia, England, North America, France, then the East Indies, Australia, and South Africa, with the task of observing celestial zones of 5° width, on average. This effort resulted in twenty catalogues appearing, one after another, beginning in 1870. They constitute what is called the AGK1. A detailed list can be found in *Traité d'astronomie stellaire* by C. André (1899). The stars in the AGK1 are designated by the letters AG, followed by the name of the observatory for the corresponding celestial zone and the number of the object in the catalogue: for example, AG Bonn 9419, AG Leipzig 742, etc. There are close to 200,000 stars, distributed over the whole sky, in the AGK1. The precision is the same as that of Lalande's catalogue. Double-star observers at the beginning of this century, particularly Aitken and Hussey, identified their discoveries by reference to Lalande's catalogue and the AGK1.

Catalogues of stars have to be remade at regular intervals, however, because the positions of stars change as a result of their proper motions and of the motions of the fundamental planes. When these motions are known exactly, it will be sufficient to extrapolate the positions taken from the catalogues; but this time will never arrive, because we become more and more exacting, and the technique of observation is constantly improving. Since 1930, therefore, the stars of the AGK1 have been reobserved with the astrograph of the Hamburg-Bergedorf observatory, with a view to a new publication that would permit determination of the proper motions of all the stars observed. Meanwhile, in 1937, the *General Catalogue of 33,342 Stars for the Epoch 1950* was published by B. Boss. It is commonly designated by the initial GC. In this work, all the stars whose positions have been measured precisely with meridian circles are collected. Despite its title, it is not a general catalogue. Its greatest value, since its appearance, has been the provision of a large number of proper motions, which has permitted the elaboration of modern stellar dynamics.

Needing to know the proper motions of many objects, Frank Schlesinger and Ida Barney of Yale Observatory reobserved the AGK1 stars from 1928 onwards, with a wide-angle astrograph. The *Yale Catalogue,* published from 1939 to 1959, contains 137,668 stars; it gives visual and photographic magnitudes, as well as spectral types and proper motions. Nevertheless, it is not of as great precision as the GC.

The new catalogue forming a continuation of the AGK1 appeared in 1954, with the title *Zweiter Katalog der Astronomische Gesellschaft für des Aquinoktium 1950*. This is the AGK2—a homogeneous work and not a somewhat ill-assorted collection like its predecessor. It comprises 183,312 stars from declination $-2°.5$ to the North Pole. Not all the sky can be found in it, therefore, but only that part of the celestial sphere easily accessible to the Hamburg Astrograph. The positions are of the same precision as those of the AGK1, but the magnitudes are photographic. A great convenience is the references it gives to Argelander's general catalogue, which we will study shortly.

In order to determine proper motions more precisely, it was proposed to take new plates like those of the AGK2, beginning in 1958, for the compilation of an AGK3. This catalogue has just been published by W. Dieckvoss of Hamburg-Bergedorf; it is also available on magnetic tape. Stars have the same number in the AGK2 and the AGK3, which is useful. The later catalogue is more precise; positions are given to $0^s.001$ and $0''.01$, and annual proper motions to four decimals. It also contains spectral types, photographic magnitudes, and references to Argelander's catalogue. The AGK3 is a fundamental source of data for the astrometrist; it is at the same time the most complete and most accurate tool that exists.

The AGK3 and Yale catalogues cover only a part of the sky, and are insufficient for the study of the orbits of artificial satellites. It was necessary, therefore, to make a catalogue for this purpose that covered the whole sky. The Smithsonian Astrophysical Observatory undertook this task, and, in 1966, published a catalogue called the *SAO Star Catalog.* It contains, in four parts, the positions, proper motions, and spectral types of 258,997 stars over the whole sky, together with references to the AGK2 and Yale catalogues, and to Argelander's catalogue. Moreover, it combines the AGK with the Yale and the FK4 stars. It can also be used in magnetic-tape form. It is less complete than the AGK2 and 3, especially in the region of the Milky Way, but it does cover the whole sky. In particular, calculations of occultations, made with a view to the detection of new binaries, are made from this catalogue.

GENERAL CATALOGUES

General catalogues are exhaustive lists of objects down to a given magnitude. There are two general stellar catalogues: Argelander's and the one made from the *Carte du Ciel.*

Argelander's catalogue is the fundamental reference for double stars. It dates from 1850. Stars in it are designated by the symbols BD (Bonner Durchmusterung) or DM. In total, the catalogue contains 457,847 stars from the North Pole to declination $-23°$. Argelander adopted a very convenient new presentation. His catalogue is divided into zones of $1°$; in each zone the stars are arranged in order of right ascension. Thus, one writes BD $+31°$ 2152 (9.0), with the magnitude in parentheses. As there are about 5,000 stars in a zone, except near the pole, such an arrangement indicates approximately the position of the star in the sky. This presentation has been repeated in the AGK2. Argelander made his observations at Bonn with a small 72-mm refractor, in the remarkably short time of 10 years. His aim was to record all the stars down to magnitude 9, and the brighter ones between magnitudes 9 and 10. Nearly half the stars are in the latter category. Anyone who uses the BD soon finds that many of the stars in it are of magnitude 10, and even 11. The positions, for the equinox 1855, are approximate to $0^s.1$ and $0'.1$, which, unfortunately, means that the work is inadequate for the calculation of the orbits of moving objects. Nevertheless, it has the merit of recording all objects, and constitutes, on that account, a remarkable document, easy to consult and constantly used by many observers. The record of stars was extended later to the South Pole, first by Schoenfeld, Argelander's student, and then photographically at the Cape of Good Hope by the Cape Photographic Durchmusterung.

The inaccuracy of positions in Argelander's catalogue, and the growing needs of astronomy, made necessary around 1900 a new, much more extensive general catalogue. The observatory of Paris took charge of this new project, the *Carte du Ciel.* This catalogue is unwieldy, however, and not adapted to the needs of the astronomer at the eyepiece, since the number of stars exceeds 10 million and their coordinates are given linearly. The observer, therefore, always refers to Argelander's catalogue (the BD) to identify stars during the night, and to avoid losing them in the sky.

Method of Locating BD Stars

Nothing looks more like a star of magnitude 9.5 than another star of magnitude 9.5. There are hundreds of thousands of them. How can an astronomer at the powerful eyepiece of a large refractor find his way around?

At the beginning of the night he should know the precession, which can be estimated by setting on a bright star. Precession varies regularly during the course of the night; its value in right ascension is almost constant, that in declination is zero at 6^h and 18^h and has maxima of opposite signs at 0^h and 12^h. Roughly speaking, at a sufficient distance from the pole the precession in right ascension is about 3 seconds a year; its greatest value in declination is some 8 arc-minutes in 25 years. If you are familiar with the value of precession, you can set on objects without making laborious corrections for the equinox.

Since the BD contains all the stars brighter than magnitude 9.5, you are sure to find in it stars of sufficient brightness. To identify a star, draw a sketch of the position of two or three stars in the field of the finder, or of the main instrument if they are close enough. Note from the chronometer the differences in time between the passage of the objects across a north-south micrometer thread, then estimate, or read on the circle, the differences in declination, and write down the estimated magnitudes. Figure 5.1 reproduces a sketch made on 24 June 1976 at the Nice 50-cm refractor to identify a new double star, BD +31° 3890. The sketch need not reproduce the relative positions of the stars faithfully, as long as the differences are noted, but it is important to indicate the direction of north. The approximate coordinates (α and δ) of one of the stars, usually the brightest, can be determined by reading the circles, and then a search is made in the catalogue for a group with the same separations, allowing for errors in the estimates. This mode of operation is quick and practical. If there is any doubt, sketch a large number of stars.

Conversely, to set with certainty on a BD star, look in the sky for the catalogue differences between neighboring stars. Such an ex-

99 Identification of Stars

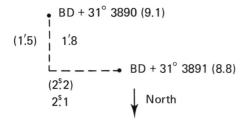

Figure 5.1 Identifying a BD star. Sketch the relative positions of the stars seen through the telescope, with estimates of their magnitudes in parentheses. The exact values, which can be found in the catalogue, make it possible to identify the star.

ercise is the first step all learners should take when they begin. Identification is sometimes difficult in the largest instruments because of flexure. For example, at Nice we sometimes have to verify with the large 76-cm refractor faint and close new double stars discovered with the 50-cm instrument. We have to begin by setting on a nearby bright star and reading the instrumental corrections before trying to find the identification sketch on the sky.

In all these operations the astronomer makes no calculations. Everything is done by circle readings, clock readings, and eye estimates. There is one chance in two to make a mistake in the signs if you make even simple calculations in the middle of the night, balancing on a ladder in the cold. Moreover, sometimes declination circles are engraved in polar distances and the numbers are then in the reverse order. The astronomer no longer bothers himself with these things after a few sessions; he makes it all automatic.

Identification of a faint star requires a good bit of patience. There is always a tendency to put aside a problem that suddenly crops up when one is observing. Many errors in the catalogues, and many lost stars, have been recognized by patient observers, whose observing records teem with accurate notes that ought always to be available for consultation. One of the first duties of observatories is to preserve these records with the greatest care.

The *Carte du Ciel*, then, was undertaken and worked out by the Paris Observatory at the beginning of the century. The sky was photographed on plates 2° square, with virtually identical astrographs, distributed in latitude. Eighteen (and later twenty) observatories participated in this task. Each observatory was equipped with an astrographic refractor of 33 cm aperture, with a focal length close to 3.44 m, giving a scale of 1 mm for 1 arc-minute. More than 12 million stars have been photographed; the positions of several millions have been reduced to rectangular coordinates.

It is outside the scope of this book to give the fundamental equations for transforming the linear coordinates on the plates into right ascension and declination. It is the classical problem of transforming from a plane to the celestial sphere, and vice versa, and is explained in treatises on wide-field astrometry. Our aim is the identification of stars, which does not require cumbersome rigorous formulae, but simple ones that can be worked out in the head. We will see from an example that a star can be identified, and its celestial coordinates obtained with the precision of the *Carte du Ciel*, by quite simple calculations.

In fact, the problem is, knowing the right ascension and declination, to find in the folios of the *Carte* a star too faint to be in the BD Catalogue. This problem is treated very well in the *Journal des Observateurs* for February 1928 by I. Lagarde. As we work within a radius of a few arc-minutes, with a "small field," we can simplify the solution still further and make it very quick by five steps:

1. Sketch the field of the star to be identified. The sketch must contain at least two stars: the one in question and a nearby one bright enough to be in the BD Catalogue or, better, the AGK2.

2. Look in the catalogues for the celestial coordinates of the bright star, in order to find out in which zone and on which plate of the zone it is. It is best to choose the equinox of 1900 for the calculations.

3. The approximate rectangular coordinates X and Y of the star for which one knows α and δ are

$$X = \tfrac{1}{4}(\alpha - A)\cos\delta, \qquad (5.1)$$
$$Y = \delta - D,$$

where A and D are the coordinates of the theoretical center of the plate. The differences $\alpha - A$ are expressed in seconds of time, and $\delta - D$ in arc-minutes. In equations (5.1) the difference between the tangent plane and the sphere is ignored, but this does not introduce any error appreciable enough to prevent identification.

4. The coordinates x and y of the catalogue are related to X and Y by linear equations of the form

$$X = K(x + Ax + By + C),$$
$$Y = K(y + A'y + B'y + C'),$$
(5.2)

where K, A, B, C, A', B', and C' are known constants for each plate. As far as we are concerned, A, B, A', and B' can be neglected. Thus the coordinates X and Y are transformed into x and y of the catalogue with the help of table 5.1, which gives simplified transformation formulae for each of the *Carte du Ciel* astrographs.

5. The values of x and y thus calculated are very close to the catalogue values x^* and y^*, which can be found at a glance. On the other hand, the differences $\Delta\alpha$ and $\Delta\delta$ between the bright star and the one to be identified have been noted on the sketch. They can be converted by

$$\Delta x = (4/K)\cos\delta\, \Delta a,$$
$$\Delta y = \Delta\delta/K,$$
(5.3)

which gives the approximate linear coordinates of the star to be identified: $x^* + \Delta x$ and $y^* + \Delta y$. In the catalogue one reads

$x^* + \Delta x^*$,
$y^* + \Delta y^*$.

The exact differences Δx^* and Δy^* give the values of $\Delta\alpha^*$ and $\Delta\delta^*$ that must be applied to the celestial coordinates α and δ of the bright star, which are already known.

Example

The example chosen has been extracted from the observing record of the 50-cm refractor for 16 November 1971. While looking at the star AGK + 23° 586 (11.2) = BD +23° 1147 (9.5) we noticed an unknown pair in the field, too faint to be in the BD. We wish to find it in the *Carte du Ciel*.

1. A sketch of the field was made, with the help of the chronometer (see fig. 5.2). The rest could be done the next day in the office.

Table 5.1 Transformation Formulae for *Carte du Ciel*.

Observatory	Zone Allocated	Simplified Transformation Formulae[a]	
		$x =$	$y =$
Greenwich	+65° to +90°	$0.2X - c + 14$	$0.2Y - f + 14$
Vatican	+55° to +64°	$0.2X + 13 + C$	$0.2Y + 13 + F$
Catania	+47° to +54°	$X - c$	$Y - c'$
Helsingfors	+40° to +46°	X	Y [b]
Hyderabad	+36° to +39°	$0.2X + 13 + C$	$0.2Y + 13 + F$
Uccle-Paris	+34° and +35°	X	Y
Potsdam-Oxford	+32° and +33°	$0.2X + 13 + C$	$0.2Y + 13 + F$
Oxford	+25° to +31°	$0.2X + 13 + C$	$0.2Y + 13 + F$
Paris	+18° to +24°	X	Y [c]
Bordeaux	+11° to +17°	$X - A$	$Y - A'$
Toulouse	+5° to +10°	X	Y
Alger	-2° to +4°	X	Y
San Fernando	-3° to -9°	$X - c_x$	$Y - c_y$
Tacubaya	-10° to -16°	$0.2X - c$	$0.2Y - c$
Hyderabad	-17° to -23°	$0.2X + 13 + C$	$0.2Y + 13 + F$
Cordoba	-24° to -31°	$0.2X + 13 + C$	$0.2Y + 13 + F$
Perth	-32° to -40°	$0.2X + 13 + C$	$0.2Y + 13 + F$
Cape	-41° to -51°	X	Y
Sydney	-52° to -64°	$-0.2X + 14 + C$	$0.2Y + 43 + F$
Melbourne	-65° to -90°	$X - c$	$Y - f$

a. Constant term of these formulae can be found at the top of each plate in the *Carte du Ciel*.
b. The catalogue gives α and δ for 1900.
c. Values of α and δ have been published for stars brighter than photographic magnitude 10.

103 Identification of Stars

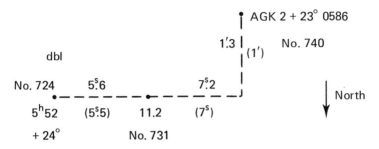

Figure 5.2 Identifying a faint star. The observed magnitudes are entered in parentheses, and should be compared with those given in the Carte du Ciel.

2. The 1900 coordinates of the star AGK +23° 586 were calculated:

$\alpha = 5^h 55^m 00^s.9$,
$\delta = +23°54'50''$.

The star should be sought, therefore, in the Paris zone on the plate centered on $D = +24°$, $A = 5^h 52^m$.

3. The approximate linear coordinates were calculated:

$X = 0.25(\alpha - A)\cos\delta = 41.35$,
$Y = \delta - D = -5.2$.

4. For Paris, the transformation formulae are $x = X$ and $y = Y$, and the true values are found from the catalogue:

$x^* = 41.51$,
$y^* = -5.13$.

It is the star no. 740 on the plate, of magnitude 11.0.

5. The new double is identified:

$\Delta x = -(12.5\cos\delta)/4 = -2.86$,
$\Delta y = +1$,

whence

$x^* + \Delta x = 41.51 - 2.86 = 38.65$,
$y^* + \Delta y = -5.13 + 1 = -4.13$.

104 Identification of Stars

Vent jeudi 2 octobre 1975 = 1975,752   anneaux peu aplatis
moins bons au milieu
de la nuit.

COU 1161   19.28.6 + 28.10
8.5 . 8.5
− 30
  15.2   8me   très douteux
  17.2 1°pos   8.7 - 8.7   4
  22.8   18.3   cp. gênant
  18.1   18:   0"14

COU 1324   20.26.5 + 31.08
9.8 - 10.2
− 1.20
                          9.4 - 9.6   ø
  36.0   8me
  38.0 3°Q   4 + Bif.
  41.4
  43.7 40,46
  43.2
  202.3 220,5   0"2,8

COU 1318   20.04.3 + 31.39
9.3 - 9.3
− .55
                          9.1 - 9.1   doit bien doubler
  89.6   Difficile 8me
  87.1 1°pos   4.p
  92.7   89.5
  88.6   90:   0"14

Examen AGK.2   21.21.7

AGK2 + 31.2019   4+   AGK2 + 31.2023   4+
       31.2020   31.2024
       29.2321   30.2123   ⟨8590 x
       29.2322   30.2124   179 k⟩
       30.2121   31.2025
       31.2021   31.2026
       31.2022   31.2027
       19.2323   19.2326
       30.2122   19.2327
       19.2324   30.2125 d(ci-dessous)
       29.2325   30.2126

AGK2 + 30.2125 (10.8) = **COU 1337**   20.23.6 + 30.12

  26.0   707 195 972   10.0 - 10.0
  13.4 3°Q   205.0   817 182 977   dbl ⋅⋅⋅ 3
  25.8   25.0   818 181 979   1'7
  24.6   ─────   3°Q
                      186 ,"56

Identification of Stars

From the catalogue we find

$x^* + \Delta x^* = 38.58,$
$y^* + \Delta y^* = -3.89,$

that is, star no. 724 of magnitude 10.3. The new double star, therefore, is

Paris +24° 5^h52 No. 724 (10.3).

Sometimes the name of the observatory is replaced by the initials AC (Astrographic Catalogue). At the same time we noted that the small star midway between the others, marked as of magnitude 11.2 on the sketch, is no. 731 (10.8).

It remains to find the exact celestial coordinates of the new double star. The true differences are obtained from the catalogue:

$\Delta \alpha^* = \Delta x K/(4 \cos \delta) = 12\overset{s}{.}8,$
$\Delta \delta^* = K \Delta y = +1\overset{\prime}{.}3,$

and the position of the new double is

$\alpha = 5^h 54^m 48\overset{s}{.}1, \quad \delta = +23°56'08''$ (equinox 1900).

Since the plates of the *Carte du Ciel* overlap, a star is almost always on two of them. Choose the plate on which it is closest to the center.

Figure 5.3 A page from the observing book of the 50-cm refractor. Three COU pairs are identified by their provisional numbers (e.g., COU 1161), their magnitudes at the time of discovery (e.g., 8.5–8.5), and their celestial coordinates (in the middle of the line). Under the magnitudes, their distances from the meridian hours and minutes (or hour angles) are given. Then a series of four or five measures of position angle and their mean can be read, together with a note of the quadrant. The separations are estimated because the pairs are very close: $0\overset{\prime\prime}{.}14$–$0\overset{\prime\prime}{.}28$. There are other notes at the end of the line: the eyepiece used (8 mm this time, which corresponds to a magnification of 938×), the estimated magnitudes, an estimate of the seeing (4–4.5, which corresponds to very good images on the 0–5 scale, and some short remarks. On the second part of the page the results of an examination of stars in the AGK2 are recorded. Of 22 stars examined, one new double is recorded in the following measure. Good identification is ensured by referring the double to a nearby star. The observations made on this night covered 11 pages and totaled 32 measures. Among 407 stars examined, 9 pairs were discovered.

Finally, to make easier the work of those who will study the star, it is usual to indicate the difference in right ascension in seconds, and of declination in arc-minutes, from the nearest BD star. The complete identification of the new binary is then

Paris +24° 5ʰ52 No. 724 (10.3)
5ʰ57ᵐ9 +23°56' (1950)
at −13ˢ and +1' from BD +23° 1147 (9.5).

SPECIAL CATALOGUES

As their name indicates, these are lists of stars that were observed for a particular purpose. There are numerous special catalogues; we will consider only those that are very useful to double-star observers, who are especially interested in pairs in orbital motion, that is, pairs of large parallax. They look for criteria of proximity that can be found in the catalogues of trigonometrical parallaxes, proper motions, and individual spectral types. Parallaxes and proper motions for nearly 6,000 stars are to be found in the *General Catalogue of Trigonometrical Parallaxes* by L. Jenkins (1952) and its 1963 supplement. W. Gliese's *Catalogue of Nearby Stars* (1969) gives parallaxes, proper motions, and a host of notes. It is of more use to the double-star observer than Jenkins's work, which is not a catalogue of nearby stars.

In 1971, Lowell Observatory in Flagstaff, Arizona, published a catalogue of 9,000 stars of large proper motion, which also contains the magnitudes and color indices of objects. Positions of stars are given to the nearest second, and to 0.1 arc-minute in declination, with references as possible to the Yale and Argelander catalogues. This catalogue is a result of the hunt for red dwarfs, vigorously pursued by, among others, W. J. Luyten and H. L. Giclas. This important work contains a reserve of many unknown pairs, but their discovery will need powerful objectives, since the objects are faint. The Lowell catalogue provided the starting point for the research program on trigonometric parallaxes of faint stars undertaken by K. Aa. Strand with the astrometric reflector of the U.S. Naval Observatory. Already, 200 parallaxes of faint dwarf stars have been published in the last few years.

A catalogue of faint red stars was issued in 1947 by Dearborn Observatory. Despite the title, *Survey of Faint Red Stars,* this is of fairly bright stars, accessible visually with average instruments. Nearly 20,000 objects are to be found in this work. Their spectroscopic characteristic is the visibility of the absorption bands of titanium oxide. Unfortunately, the positions lack precision, and no references are given to other catalogues.

To conclude the list of special catalogues, we should mention the *Henry Draper Catalogue* (HD for short), which gives the spectra of 222,000 stars distributed all over the sky. This important work, completed by Harvard College Observatory at Cambridge, Massachusetts, and Arequipa, Peru, was published between 1918 and 1924. The HD catalogue serves as a reference for stellar spectrograms. Fortunately, since the HD positions are imprecise, it gives the corresponding BD numbers, which permit identification. Double-star observers never identify their stars by reference to the HD (which is contained, for all practical purposes, in the AGK3 and the SAO), but "spectroscopists" seem to ignore these latter catalogues. They nearly always observe bright stars whose identification is obvious. All the same, the astrometrists have to check these objects.

As a general rule, a reference to a catalogue should always be given for a star, as well as its position and magnitude. Observers lose precious time recovering the coordinates of stars that have been noted casually. It can happen that objects are taken without comment from little-known catalogues, indicated by indecipherable initials, with neither position nor magnitude. Moreover, those who name the object do not always know the significance of the abbreviations used by their predecessors, which they just reproduce.

BIBLIOGRAPHY

Fundamental Catalogues
The history of fundamental catalogues can be found in the most recent of them:

Fricke, W., and A. Kopff. Fourth fundamental catalogue FK4. *Veröffentlichungen Astronomisches Rechen-Institut,* Heidelberg, no. 10 (1963).

A large bibliography is given in the following:

Scott, F. P. "The Systems of Fundamental Proper Motions." In *Basic Astronomical Data.* University of Chicago Press, 1963.

108 Identification of Stars

Intermediate Catalogues

The history of catalogues before 1900 is described in the following:

André, C. *Traité d'astronomie stellaire.* Paris: Gauthier-Villars, 1899.

Boss, B. *General Catalogue of 33,342 Stars for the Epoch 1950.* Washington, D.C.: Carnegie Institution, 1937.

The AGK2 is

Schorr, R., and A. Kohlschütter. *Zweiter Katalog der Astronomischen Gesellschaft für das Aquinoktium 1950.* Hamburg-Bergerdorf, 1954.

The Yale Catalogue was published in twenty fascicules of the *Transactions of the Astronomical Observatory of Yale University* between 1939 and 1959. The SAO Star Catalog was published by the Smithsonian Astrophysical Observatory, Cambridge, Massachusetts, in 1966.

General Catalogues

Argelander's catalogue was reissued in 1903:

Argelander, F. *Bonner Durchmusterung des Nördlichen Himmels.* Bonn, 1903.

The list of observatories that have contributed to the *Carte du Ciel* can be found in the text. Some results were published only after the second world war, namely those for the zones +32 and +33 (in 1953), +34 (in 1960), +35 (in 1962), and +36 (in 1946). Note two valuable works: *Catalogue photographique du ciel.* Paris: Gauthier-Villars, 1946. (Right ascensions and declinations deduced from the rectilinear coordinates for all stars down to magnitude 9.9 (equinox 1900). Zones +18°, 20°, 22°, 24°. Paris Observatory.) *Catalogues de 11,755 etoiles de la zone +17° a +25° et de magnitude 9.5 a 10.5 destinées à servir de références pour la détermination des mouvements propres des étoiles du catalogue photographique de Paris.* Paris Observatory, 1950.

Van Biesbroeck, G. "Star Catalogues and Charts." In *Basic Astronomical Data.* University of Chicago Press, 1963. (This article gives useful tables for calculating the positions of stars from the rectilinear coordinates of the *Carte du Ciel* plates.)

Heckmann, O., W. Dieckvoss, and H. Kox. "New Plate Constants in the System of the FK 3 for the Declination Zones +21°, +22°, +23°, +24° of the Astrographic Catalogue, Paris." *Astronomical Journal* 49 (1954): 143.

Special Catalogues

Aravamudan, S. "Stars with Large Proper Motions in the Astrographic Zones +32° and +33°." *Journal des Observateurs* 42 (1959): 123.

Blanco, V. M., S. Demers, G. G. and M. P. FitzGerald. Photometric Catalogue. Magnitudes and colors in the U,B,V and U_c,B,V systems. Publications of the

U.S. Naval Observatory, second series, vol. 21 (1970). (There are 20,705 objects in this catalogue, including many of the classical double stars. Spectral types and luminosity classes are given.)

Cannon, A. J., and E. C. Pickering. The Henry Draper Catalogue. *Annals of the Harvard Observatory,* vols. 91 ff. (1918).

Giclas, H. L., R. Burnham, Jr., and N. G. Thomas. *Lowell proper motion survey: Northern Hemisphere.* The G-numbered stars. 8,991 stars fainter than magnitude 8 with motions >0."26/year. Flagstaff, Arizona: Lowell Observatory, 1971.

Gliese, W. Catalogue of Nearby Stars. Institut Heidelberg report no. 22 (1969).

Hoffleit, D. *Catalogue of Bright Stars.* New Haven, Conn.: Yale University, 1964.

Jenkins, L. F. *General Catalogue of Trigonometric Stellar Parallaxes.* New Haven, Conn.: Yale University, 1952.

———. *Supplement to the General Catalogue of Trigonometric Parallaxes.* New Haven, Conn.: Yale University, 1963.

Klemola, A. R., S. Vasilevskis, C. D. Shane, and C. A. Wirtanen. Catalogue of proper motions of 8,790 stars with reference to galaxies. Publications of Lick Observatory, vol. 22, part II (1971).

Lee, D. U., R. J. Baldwin, T. J. Bartlett, G. D. Gore, and D. W. Hamlin. Dearborn Catalog of Faint Red Stars: Titanium Oxide Stars. *Annals of Dearborn Observatory* 5 (1943): part 1A; (1944): part 1B; 1947: part 1C.

Riddle, R. K. First Catalog of Trigonometric Parallaxes of Faint Stars: Astrometric Results. Publications of the U.S. Naval Observatory, second series, vol. 20 (1970): part 3.

Routly, P. M. Second Catalog of Trigonometric Parallaxes of Faint Stars. Publications of the U.S. Naval Observatory, second series, vol. 20 (1972): part VI.

Vyssotsky, A. N. "Dwarf M stars found spectrophotometrically." *Astronomical Journal* 61 (1956): no. 5.

There are references to other lists of 876 objects in all. C. E. Worley has discovered 30 close pairs in these lists.

6
COMPUTATION OF ORBITS AND STELLAR MASSES

THE AIM OF DOUBLE-STAR OBSERVATION

The aim of observing double stars is to determine their orbits, which, in turn, lead to their masses if their distances from the solar system are known. These masses are obtained from the observed motions, and are fundamental or geometric masses; they are not based on any hypothesis except the validity of the law and the constant of gravitation. All other methods (for example, those by which so-called photometric or spectroscopic masses are obtained) appeal to more or less plausible hypotheses about stellar characteristics. We shall see that the study of double stars enables us to obtain so-called dynamic masses, and also to determine, in certain cases, the absolute parallax, and thus the masses, by measuring the radial velocities. In this chapter, after recalling the basic concepts needed for the calculation of orbits, we will review some of the methods and then apply them to the calculation of absolute and dynamic masses. Finally, we will draw up a balance sheet of our knowledge after two centuries of observation.

THE TRUE ORBIT

The secondary or companion star B moves with respect to the primary star A as if it were a fixed center of attraction having a mass equal to the whole mass of the system.

Let an ellipse have its center at D and one focus at A. The periastron P is the end of the major axis nearest to A; the other end is the apastron. The elements or parameters of the true orbit are the following:

Computation of Orbits and Stellar Masses

- the period of revolution of the companion in years, P,
- the mean annual motion, $n = 360/P$ or $\mu = 2\pi/P$,
- the time of periastron passage, T,
- the orbital eccentricity, e (sometimes we write $e = \sin\phi$), and
- the major semiaxis, a (expressed in angular measure or astronomical units).

A theorem of celestial mechanics shows that the trajectory of B, or its orbit, is an ellipse, one of whose foci is occupied by the primary star A. The proof can be found in specialized treatises; this chapter will emphasize what is necessary for the computation of orbits, and assume that the reader knows the elementary geometric properties of conics.

We wish to determine the position, in its orbit, of the secondary star B, as a function of time. The position is defined by two parameters: the radius vector $AB = r$ and the angle $v = \angle PAB$, called the *true anomaly*. Draw the circumference of a circle of radius OP (fig. 6.1). This circle is called the *auxiliary circle*. Drop the perpendicular BB' on OP; it intersects OP at Q. The angle $\angle POB = u$ is called the *eccentric anomaly*. Take the direction along OP, from O towards P, as positive; then

$$AQ = AO + OQ = -ae + a\cos u = r\cos v, \tag{6.1}$$

$$QB = QB'(1 - e^2)^{1/2} = a(1 - e^2)^{1/2}\sin u = r\sin v, \tag{6.2}$$

since the ellipse is the projection of the auxiliary circle from a plane inclined at an angle ϕ. From equations (6.1) and (6.2) we deduce

$$r = \frac{a(1 - e^2)}{1 + e\cos v}. \tag{6.3}$$

We still have to express the true anomaly v in terms of the time t. This is done through the eccentric anomaly u. We know that the radius vector sweeps out equal areas in equal times:

$$\frac{1}{2}r^2\frac{dv}{dt} = c = \frac{\pi a^2 \cos\phi}{P} = \frac{\mu a^2 \cos\phi}{2},$$

where c is the areal constant, equal to the area of the ellipse divided by the period.

112 Computation of Orbits and Stellar Masses

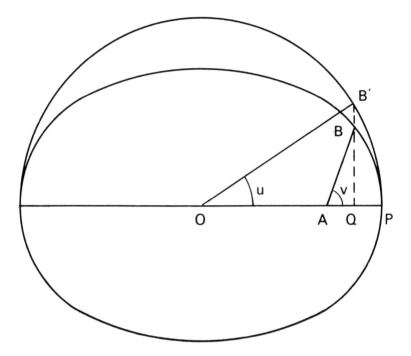

Figure 6.1 The true orbit.

On the other hand,

$$\tan(v/2) = \left(\frac{1+e}{1-e}\right)^{1/2} \tan(u/2), \tag{6.4}$$

which, by differentiation, gives

$$\frac{du}{dt} = \frac{\mu}{1 - e\cos u},$$

from which, by integration,

$$u - e\sin u = \mu(t - T) = M.$$

It is usual to express M in degrees; then,

$$M = n(t - T). \tag{6.5}$$

The angle *M* is called the *mean anomaly*. It is proportional to the time reckoned from the periastron passage. It corresponds to the mean position of the companion, whereas *v* corresponds to the true position. The difference $v - M$ is called the *equation of the center*. Tables calculated by means of equations (6.4) and (6.5) make it possible to convert *M* into *v*.

The quantities

$$X = (r/a)\cos v = \cos u - e,$$
$$Y = (r/a)\sin v = (1 - e^2)^{1/2}\sin u \qquad (6.6)$$

are called the reduced coordinates of the component *B*. There are tables giving these quantities for all values of *M* as a function of eccentricity.

THE APPARENT ORBIT

The apparent orbit is the projection of the true ellipse onto the plane of the sky. To the elements of the true orbit, called *dynamic elements,* the geometric elements that define its orientation with respect to the plane of the sky must be added. Consider figure 6.2, in which the plane of the true orbit is seen projected onto that of the sky. The intersection of the two planes is NN'. The primary star *A* is in both planes. The companion *B'* is projected onto *B*. The major semiaxis of the true orbit is expressed in arc-seconds. There are two kinds of geometric elements: Campbell's and the Thiele-Innes elements.

Campbell's Elements

These are the following:
- The inclination *i* between the plane of the orbit and the tangent plane to the celestial sphere. The inclination lies between 0° and 180°; it is less than 90° if the orbital motion is direct.
- The position angle Ω of the intersection NN'. This intersection is called the *line of nodes*. The position angle is reckoned from an arbitrary origin, the direction of the North Pole. The line of nodes cuts the ellipse at two points: *N* is the descending node corresponding to a radial velocity of approach; *N'* is the ascending node. In

114 Computation of Orbits and Stellar Masses

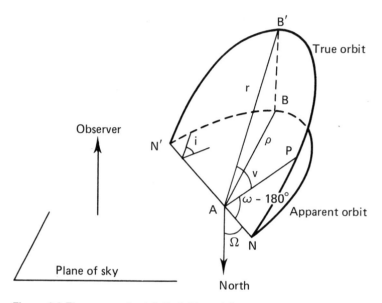

Figure 6.2 The apparent orbit. Definition of the parameters.

general we do not know which of the nodes is the ascending one, so, from the two possible values of Ω, we choose the one less than 180°

• The angle ω between the ascending node and the periastron, measured in the plane of the true orbit and in the direction of orbital motion.

The following table summarizes the elements of the orbit of a visual double star.

Dynamical Elements

P period in years

n mean motion in degrees per year, $360/P$

T time of periastron passage

e eccentricity

a major semiaxis

---

115 Computation of Orbits and Stellar Masses

Campbell Geometric Elements

i inclination

Ω position angle of the node

ω angle between the node and periastron

The derivation of these elements from the observations determines the orbit of a double star.

We are going to derive the transformation formulae between these elements and the observed quantities already studied in chapter 3, namely, the separation ρ, the position angle θ, and the time t.

According to the above definitions, from figure 6.2 we obtain, for direct motion,

$AB = \rho$, $AB' = r$,

$\angle PAB = v$, $\angle NAP = \omega$, $\angle BAN = \theta - \Omega$.

Project ρ and r onto the line of nodes NN', and then onto the perpendicular to that line. We have, first,

$$\rho \cos(\theta - \Omega) = r \cos(v + \omega); \tag{6.7}$$

then

$$\rho \sin(\theta - \Omega) = r \sin(v + \omega) \cos i, \tag{6.8}$$

whence the fundamental equations

$$\tan(\theta - \Omega) = \tan(v + \omega) \cos i,$$

$$\rho = \frac{r \cos(v + \omega)}{\cos(\theta - \Omega)} = \frac{a(1 - e^2)}{1 + e \cos v} \frac{\cos(v + \omega)}{\cos(\theta - \Omega)}, \tag{6.9}$$

$$M = n(t - T).$$

We add to these equations the one that gives the radial component:

$$BB' = z = r \sin(v + \omega) \sin i. \tag{6.10}$$

From equations (6.9), an ephemeris can be computed for a double star; that is, a table giving ρ and θ at time intervals that may or may not be regular.

In practice, one needs tables of Keplerian motion that give the true anomaly v, or better the equation of the center $v - M$, as a function of the mean anomaly M, for all values of the eccentricity. Such a table can be found in Danjon's *L'Astronomie générale* (p. 432). The U.S.S.R. Academy of Sciences published a very complete table in 1960, which, unfortunately, is not often seen outside astronomical institutes. With pocket electronic calculators, an ephemeris can be computed directly. An example is given in table 6.1.

The Thiele-Innes Elements

We choose a three-dimensional coordinate system having for origin the primary star A (fig. 6.3). The axis Ax is directed towards celestial north, the axis Ay towards increasing right ascensions, and the axis Az towards the observer. The companion B' is projected onto the sky at B. We have $AB = \rho$, $\angle xAB = \theta$, $AB' = r$. The coordinates of B' are

$$x = \rho \cos\theta,$$
$$y = \rho \sin\theta, \tag{6.11}$$

and z given by equation (6.10). With the help of the reduced coordinates of equation (6.6), we find after some elementary calculations

$$x = AX + FY,$$
$$y = BX + GY, \tag{6.12}$$
$$z = CX + HY,$$

where

$$A = a(\cos\omega \cos\Omega - \sin\omega \sin\Omega \cos i),$$
$$B = a(\cos\omega \sin\Omega + \sin\omega \cos\Omega \cos i),$$
$$F = a(-\sin\omega \cos\Omega - \cos\omega \sin\Omega \cos i), \tag{6.13}$$
$$G = a(-\sin\omega \sin\Omega + \cos\omega \cos\Omega \cos i),$$
$$C = a \sin\omega \sin i,$$
$$H = a \cos\omega \sin i.$$

The constants A, B, F, G, C, and H are known as the *Thiele-Innes elements*; they determine the plane of the true orbit, and its size, just as Campbell's elements ω, Ω, i, and a do.

We should also write down the equations for transforming the

117 Computation of Orbits and Stellar Masses

Table 6.1 Ephemeris for Castor.

t	$t-T$	M	v	$v+\omega$	$\tan(v+\omega)$	$\tan(\theta-\Omega)$	$\theta-\Omega$	θ
1978.0	27.35	19°.26	41°.82	281°.63	−4.8587	1.8938	62°.16	103°.8
79.0	28.35	19.96	43.21	283.02	−4.3246	1.6856	59.32	101.0
80.0	29.35	20.67	44.60	284.41	−3.8919	1.5169	56.61	98.3
81.0	30.35	21.37	45.95	285.76	−3.5434	1.3811	54.09	95.7
82.0	31.35	22.07	47.29	287.10	−3.2506	1.2670	51.72	93.4
83.0	32.35	22.78	48.63	288.44	−2.9991	1.1690	49.45	91.1
84.0	33.35	23.48	49.95	289.76	−2.7837	1.0850	47.33	89.0

t	$\cos v$	r	$\cos(v+\omega)$	$\cos(\theta-\Omega)$	$\dfrac{\cos(v+\omega)}{\cos(\theta-\Omega)}$	ρ
1978.0	0.7452	5.0572	0.2016	0.4669	0.4317	2″.18
78.0	7288	5.0808	2253	5102	4416	2.24
80.0	7120	5.1053	2489	5504	4521	2.31
81.0	6953	5.1299	2716	5865	4631	2.38
82.0	6783	5.1552	2940	6196	4746	2.45
83.0	6609	5.1812	3163	6501	4866	2.52
84.0	6435	5.2076	3381	6777	4988	2.60

Orbital elements $P = 511\overset{y}{.}3$ $n = 0°.70409$ $T = 1950.65$ $a = 7″.369$
$e = 0.36$ $i = 112°.94$ $\Omega = 41°.65$ $\omega = 239°.81$

118 Computation of Orbits and Stellar Masses

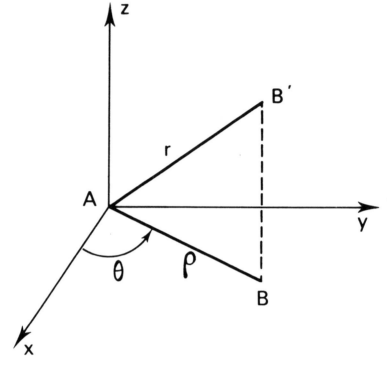

Figure 6.3 Coordinate system for Thiele-Innes elements.

Thiele-Innes elements into those of Campbell. They can easily be found from equations (6.13):

$$\tan(\Omega + \omega) = \frac{B - F}{A + G} \quad \text{with } 0° < \Omega < 180°,$$

$$\tan(\Omega - \omega) = \frac{B + F}{A - G},$$

$$\tan^2(i/2) = \frac{(B + F)\sin(\Omega + \omega)}{(B - F)\sin(\Omega - \omega)},$$

$$a^2 = \frac{AG - BF}{\cos i}.$$

(6.14)

119 Computation of Orbits and Stellar Masses

When you give the orbital elements of a double star, it is good practice to give both the Campbell elements and the Thiele-Innes elements. An ephemeris can be computed from the Thiele-Innes elements much more directly, with the help of tables of the reduced coordinates X and Y, for all values of M, as a function of the eccentricity. All the same, these tables are laborious to use, since the numbers and their differences are given to five decimals in small characters and place a great strain on the eyes. Once X and Y are obtained, equations (6.12) give directly

$\rho = (x^2 + y^2)^{1/2}$,
$\theta = \tan^{-1}(y/x)$.

The radial velocity of the secondary with respect to the primary is

$$\frac{dz}{dt} = \frac{CdX}{dt} + \frac{HdY}{dt} = \frac{nCdX}{dM} + \frac{nHdY}{dM} \quad (M \text{ in degrees}).$$

Thus, the radial velocity is given in arc-seconds per year. It can be reduced to kilometers per second by introducing the absolute parallax p and a conversion factor for the units:

$$V_{\text{kms}^{-1}} = \frac{LdX}{dM} + \frac{NdY}{dM} \tag{6.15}$$

with $L = 4.737nC/p$, $N = 4.737nH/p$. We also note that, by equation (6.10),

$$V_{\text{kms}^{-1}} = 4.737\mu \frac{a \sin i}{p \cos\phi} e \cos\omega \cos(v + \omega) \; (a \text{ and } p \text{ in arc-sec}). \tag{6.16}$$

Equations (6.15) and (6.16) show that the measurement of relative radial velocity gives an absolute parallax for the system when the orbital elements are known.

In addition, we should give some indication of the geometric significance of the constants A, B, F, and G. It can easily be seen that the coordinates of the center of the apparent ellipse are

$$x_c = -Ae, \qquad y_c = -Be. \tag{6.17}$$

If the center of the apparent ellipse is taken as the origin of new axes (ξ, η) parallel to the former ones, and of the same sense, then, for a point of true anomaly v,

$$\xi = A\cos v = F_1 \sin v,$$
$$\eta = B\cos v = G_1 \sin v, \tag{6.18}$$

where $F_1 = F\cos\phi$, $G_1 = G\cos\phi$. These equations show that A and B are the coordinates of periastron P (where $v = 0$), with respect to the center, and that F_1 and G_1 are the coordinates of the point with true anomaly $v = 90°$, which is situated on the conjugate diameter of the projection of the major axis, at the first intersection met in going from P in the direct sense. Finally, the direction cosines of the positive directions of the major and minor axes of the true orbit are

$A/a, B/a, C/a$

and

$F/a, G/a, H/a$.

All the orbital elements are summarized in table 6.2. Once these definitions are given, the computation of orbital elements becomes very simple.

COMPUTING THE ORBITAL ELEMENTS OF A DOUBLE STAR

There are several methods of orbit computation, and they fall into two categories: geometric methods and analytic methods. The first are quicker but lead to less certain values of the elements. They are useful for orbits that are poorly observed, either because the systems are difficult and observable only with powerful instruments or because the length of the period permits the observation of only a more or less significant arc of the whole trajectory. In general,

Table 6.2

Dynamical Elements	Campbell Elements	Thiele-Innes Elements	
P		A	
n	i	B	C
T	Ω	F	H
e	ω	G	
a			

121 Computation of Orbits and Stellar Masses

periods are so long that the computer does not have the patience to wait until a complete orbit has been marked out by the observations. In most cases a whole lifetime would not be sufficient. He can always recompute earlier elements that give an ephemeris that does not agree with observation, or, better, make do with a part of the apparent ellipse. The ideal case of a pair observed well throughout one revolution, whose elements have not yet been computed, is no longer to be found in the literature. If any such system exists, it is jealously kept secret by its powerfully equipped discoverer, who reserves the priority of computation for himself. Such secrets of the dome are increasingly rare, but they give the conscientious observer his just reward.

About 60 orbits are computed each year throughout the world. More than half are improvements of former orbits. Of the 700 or 800 orbits now published, only a few score have definitive elements. There is plenty of work, then, for the computer. The difficulty is the errors in the measures. Only a well-trained observer, who knows the history of the observations, is at all likely to find even approximately correct elements from disparate measures, too scarce at some epochs and too plentiful at others. There are observers whose measures cannot be faulted; others, on the contrary, can be ignored. Years of patience, however, are needed to know how to judge such-and-such an observation of a given pair. Modern methods of numerical analysis can hardly be used. The computation of the orbit of a double star will always be the work of a craftsman.

Before beginning the computation proper, all the observations should be reduced to the same equinox. In fact, the celestial pole, which serves as the origin for the position angles, is not stationary. A correction must therefore be applied to the position angles:

$$\Delta\theta = -0\overset{\prime\prime}{.}0056 \sin\alpha \sec\delta(t - t_0), \tag{6.19}$$

where t_0 is the epoch chosen for the equinox, and t the epoch of observation. Usually, t_0 is chosen to be not far from the equinox of the first observations, for example 1900 or 1950, in order to make the corrections small or negligible. The position angles become

$$\theta_0 = \theta_t + \Delta\theta. \tag{6.20}$$

Proper motion also changes the position angles. A mistake that is sometimes made is not to take this into account; the correction can be written

$$\Delta\theta = -0\overset{s}{.}00417\mu_\alpha \sin\delta(t - t_0), \tag{6.21}$$

where μ_α is expressed in seconds of time per year. This correction can become important near the pole for pairs of large proper motion. There are other corrections for the variation of parallax with time, but they are negligible. Errors committed by confusing a central projection with an orthogonal one are also negligible. One should be aware, however, that the observed period is only an apparent one. The true period is greater or less by the time that light takes to traverse the change in distance, during an orbital period, that corresponds to the radial velocity of the system. This time seldom exceeds a few hundredths of a year.

The Geometric Method

The simplest method of all is completely graphical, and enables the elements to be computed in a few minutes once the apparent ellipse has been drawn. This ellipse should be drawn with the greatest care, since the accuracy of the elements depends on the accuracy of the drawing. Several French authors have described how to do this, in particular Danjon and Baize. The individual observations, reduced to the same equinox, are plotted on a graph. If the companion has described several revolutions, a preliminary period can be determined from the time taken for it to return to the same position angles, and the observations can be grouped into one revolution. The scale chosen should be large enough to eliminate drawing errors; about 30 cm for the largest dimension is convenient.

When the observations have been plotted on the graph (the epoch and the names of the astronomers should be noted also), they form a cloud of points that often looks very different from the curve sought. For guidance, a preliminary curve $\theta(t)$ can be constructed freehand. Thus, discrepant or manifestly erroneous observations are noted; there are always some. Be careful, however, in correcting the errors thus found. The curve $\theta(t)$ traced in this way can only be approximate; its role is to indicate the sense and order of magnitude of the errors of some observations which will be treated separately,

123 Computation of Orbits and Stellar Masses

or completely eliminated. It is not advisable to construct a curve of $\rho(t)$, since measures of separation are subject to greater errors.

Once the observations have been noted, an ellipse must be traced through the cloud of points. This ellipse must satisfy the law of areas. Baize advises tracing the ellipse by the classical method of a thread and two needles. The primary star A does not occupy one of the foci of the apparent ellipse (fig. 6.4), since this latter is the projection on the sky of the true orbit. Five or six points are chosen from among the better observations; they should be distributed along the ellipse. Each point is joined to the star A, and the areas of the sectors thus defined are measured in arbitrary units with a planimeter. If S_n is the area of the sector n, and Δt_n the time taken

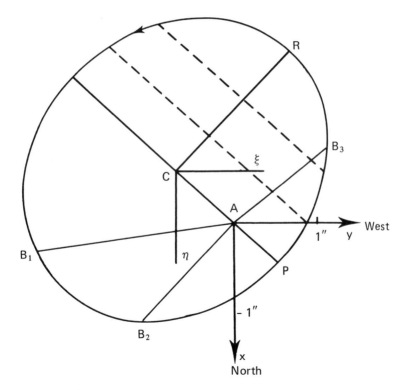

Figure 6.4 Computation of orbital elements by a geometric method.

by the companion to traverse it, the constant of areas can be written

$$c = S_n/\Delta t_n \tag{6.22}$$

and must be the same in each of the sectors chosen. To be satisfactory, the value of the constant of areas ought to vary by only a few percent from one sector to another; otherwise the elements found will not represent the observations correctly. Once the constant of areas has been found, in any units, the area of the ellipse can be measured with the planimeter in the same units. This should be done several times, and the mean (S) of the measures taken. The period then is

$$P = S/c. \tag{6.23}$$

If the period is known in advance, the two values should agree.

The computer will seldom be satisfied by the apparent ellipse he has drawn. Sometimes he has to make the ellipse pass outside the observed points. Ultimately, the process depends on the choice of the apparent foci, whose positions the computer ought to be able to modify appreciably as required. His imagination will be tested when the ellipse drawn does not obey the law of areas. It can happen that the ellipse suggested by the scatter of the observed points is grossly in error.

Once the apparent orbit has been obtained, its center (C) is carefully determined (fig. 6.4); by definition it coincides with that of the true orbit. The axis CA, the projection of the major axis, intersects the ellipse in P, the projection of the periastron. The eccentricity is then measured directly:

$$e = CA/CP.$$

The date of periastron passage can be obtained without difficulty, either by reading the epoch corresponding to the position P from the curve of $\theta(t)$ or, preferably, by using the observations selected for the determination of the constant of areas in the following manner: Consider the case of figure 6.4. Successive positions of the companion are B_1, B_2, P, and B_3. According to the law of areas,

$$T = t_{B1} + \frac{\text{area } B_1AP}{c}$$

$$= t_{B2} + \frac{\text{area } B_2AP}{c}$$

$$= t_{B3} - \frac{\text{area } B_3AP}{c}. \tag{6.24}$$

These three values of T should be very close to each other, and their mean can be taken. Moreover, as many observations as one wishes can be taken to compute T. Thus, all the dynamical elements except a have been obtained: P, $n = 360/P$ or $\mu = 2\pi/P$, T, and e.

It remains to find the elements of the orbital plane and the major semiaxis. Equations (6.18) answer this need. The coordinates of P, with respect to the center C, are A and B, and can be read from the drawing, in which the unit of length is the arc-second. In order to find the diameter conjugate to CP, draw chords parallel to CP; their midpoints define the locus sought. Let R be its intersection with the ellipse (fig. 6.4). The coordinates of the locus with respect to the center are read from the drawing:

$$\begin{aligned}\xi_R &= F_1 = F\cos\phi, \\ \eta_R &= G_1 = G\cos\phi.\end{aligned} \tag{6.25}$$

The four constants A, B, F, and G having been obtained, the conversion formulae (6.14) are used to deduce i, Ω, ω, and a, and thus the two last constants C and H used to calculate the radial velocities. It remains to compare the ephemeris given by the elements with the observations.

The Method of Thiele, Innes, and van den Bos

This method makes use of any three positions of the secondary; it is similar in principle to that used to compute the orbit of a planet. From this point of view it has great pedagogic value.

Once an apparent ellipse that faithfully obeys the law of areas has been constructed, three positions (called mean or normal places) can be chosen, in the sense of increasing time, on portions of the ellipse particularly well covered by the observations. Let

$$\begin{aligned}&t_1, \rho_1, \theta_1, &&x_1 = \rho_1 \cos\theta_1, &&y_1 = \rho_1 \sin\theta_1, \\ &t_2, \rho_2, \theta_2, &&x_2 = \rho_2 \cos\theta_2, &&y_2 = \rho_2 \sin\theta_2, \\ &t_3, \rho_3, \theta_3, &&x_3 = \rho_3 \cos\theta_3, &&y_3 = \rho_3 \sin\theta_3\end{aligned} \tag{6.26}$$

be the epochs, separations, position angles, and Cartesian coordinates for these three points. The constant of areas, c, is known from the drawing of the apparent ellipse: If S is the total area of the ellipse, the period is given by S/c. On the other hand, $\mu = 2\pi/P$ or $n = 360/P$, in radians or degrees per year, respectively.

Let us consider the triangle $A12$ (fig. 6.5). We have

$$S_{12} = \tfrac{1}{2} \rho_1 \rho_2 \sin(\theta_2 - \theta_1)$$
$$= x_1 y_2 - x_2 y_1,$$

which gives, by the equations $x = AX + FY$ and $y = BX + GY$,

$$S_{12} = \frac{c}{2\mu} [\sin(u_2 - u_1) - \sin\phi(\sin u_2 - \sin u_1)]. \tag{6.27}$$

Now, from Kepler's equation,

$$t_2 - t_1 = \frac{1}{\mu} [u_2 - u_1 - e(\sin u_2 - \sin u_1)],$$

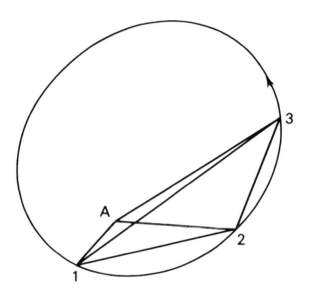

Figure 6.5 Theory of the method of Thiele, Innes, and van den Bos.

and, by simple transformations,

$$t_2 - t_1 - \frac{2S_{12}}{c} = \frac{1}{\mu}[u_2 - u_1 - \sin(u_2 - u_1)]. \tag{6.28}$$

This is Thiele's fundamental equation. Let us set $u_2 - u_1 = \alpha$, $u_3 - u_2 = \beta$, and $u_3 - u_1 = \alpha + \beta$. Then

$$\begin{aligned}\alpha - \sin\alpha &= \mu(t_2 - t_1) - 2\pi S_{12}/S,\\ \beta - \sin\beta &= \mu(t_3 - t_2) - 2\pi S_{32}/S,\\ (\alpha + \beta) - \sin(\alpha + \beta) &= \mu(t_3 - t_1) - 2\pi S_{31}/S.\end{aligned} \tag{6.29}$$

The second term of these equations is known from the ellipse that has been drawn and the observations. The first two equations give α and β; the third serves as a check. If there is any disagreement, the cause must be the way in which the apparent ellipse was drawn; in this case it is necessary to begin again from the beginning.

From the relation (6.27) are obtained, first,

$$S_{12} + S_{23} - S_{13} = \frac{c}{2\mu}[\sin\alpha + \sin\beta - \sin(\alpha + \beta)], \tag{6.30}$$

and then

$$S_{23}\sin\alpha - S_{12}\sin\beta = \frac{ce}{2\mu}[\sin\beta(\sin u_2 - \sin u_1) - \sin\alpha(\sin u_3 - \sin u_2)].$$

By $u_1 = u_2 + \alpha$ and $u_3 = u_2 + \beta$, this last relation gives

$$S_{23}\sin\alpha - S_{12}\sin\beta = \frac{ce}{2\mu}\sin u_2[\sin\alpha + \sin\beta - \sin(\alpha + \beta)]. \tag{6.31}$$

In the same way we obtain

$$S_{23}\cos\alpha + S_{12}\cos\beta - S_{13} = \frac{ce}{2\mu}\cos u_2[\sin\alpha + \sin\beta - \sin(\alpha + \beta)]. \tag{6.32}$$

The equations (6.30)–(6.32) give the two fundamental relations

$$\begin{aligned}e\sin u_2 &= \frac{S_{23}\sin\alpha - S_{12}\sin\beta}{S_{12} + S_{23} - S_{13}},\\ e\cos u_2 &= \frac{S_{23}\cos\alpha + S_{12}\cos\beta - S_{13}}{S_{12} + S_{23} - S_{13}}.\end{aligned} \tag{6.33}$$

The right-hand sides are completely known, so e and u_2 can be found; then $u_1 = u_2 - \alpha$ and $u_3 = u_2 + \beta$, and $M_1 = u_1 - e\sin u_1$, M_2, and M_3 can be calculated, from which the time of periastron passage,

$$T = t_1 - M_1/n = t_2 - M_2/n = t_3 - M_3/n,$$

is obtained. These three values should be very close.

This first step has provided the dynamical elements P, n, e, and T. It remains to compute the constants A, B, F, and G. According to (6.12),

$$x_i = AX_i + FY_i, \qquad y_i = BX_i + GY_i \qquad (i = 1, 2, 3).$$

There are six equations for four unknowns. They can be solved by least squares, or, alternatively, the unknowns can be calculated from positions 1 and 3, 2 and 1 being used as a check.

If the right-hand sides of equations (6.33) are zero, then the eccentricity is zero also: The true orbit is a circle and there is no periastron. The convention is to take the epoch of nodal passage as the origin for time. If, in addition, the inclination is zero, there is neither periastron nor node; in this case the secondary describes a circular orbit with uniform motion. Once the constants A, B, F, and G have been obtained, the elements a, i, ω, and Ω follow from equations (6.14).

This method of Thiele, Innes, and van den Bos is satisfying to the mind, but not always to the computer. The set of equations (6.33) contains differences in both the numerators and the denominators of the right-hand side, and these differences may become vanishingly small. In this case, the eccentricity and the anomaly u_2 are poorly determined. It is necessary, therefore, to begin with adequately large sectors. Everything depends on the choice of normal places, and the flair to make that correctly is acquired by experience.

Example

We choose the star ADS 1227 = A 1913, mag. 9.5–9.5, α (1950) $1^h31\overset{m}{.}6$, $\delta(1950)$ +34°24'. This star, discovered in 1908, has traversed only 120° since its discovery. In 1975, an orbit was computed by Erceg using a set of observations up to 1961. A measure made at Nice in 1976 shows that this orbit does not represent the currently

129 Computation of Orbits and Stellar Masses

observed motion. It is a good case for the application of the method of Thiele, Innes, and van den Bos.

To begin, the observations are grouped as in table 6.3. The names of the observers, with the instruments used, as well as the numbers of nights, are duly noted. In this case, the corrections for precession are negligible. The measures are plotted on a sheet of graph paper, and an ellipse is drawn (fig. 6.6). Care must be taken to respect the law of areas. After some difficulty in representing this law in the interval 1920–1948, an ellipse was found from which the values in table 6.4 are derived. If one takes into account the lack of precision of the observations, and their small number, this table can be considered satisfactory. From it we can deduce $c = 0.01757$. The area S of the ellipse is measured, whence the period $P = S/c = 2.7607/c = 157.1$.

For the normal places, the extreme observations and that of 1949 are chosen:

$t_1 = 1908.87, \quad \rho_1 = 0''.22, \quad \theta_1 = 267°.0,$
$t_2 = 1948.78, \quad \rho_2 = 0''.18, \quad \theta_2 = 205°.8,$
$t_3 = 1976.81, \quad \rho_3 = 0''.20, \quad \theta_3 = 144°.6.$

From the graph we find that

$2\pi S_{12}/S = 1.04397,$
$2\pi S_{32}/S = 1.01962,$
$2\pi S_{31}/S = 0.97114,$

which leads to

$\alpha - \sin\alpha = 0.55203,$
$\beta - \sin\beta = 0.10130$
$(\alpha + \beta) - \sin(\alpha + \beta) = 1.74578,$

$\alpha = 88°.91,$
$\beta = 49°.13, \quad \alpha + \beta = 138°.04,$
$\alpha + \beta = 138°.21.$

We adopt the values

$\alpha = 89°.0 = u_2 - u_1,$
$\beta = 49°.2 = u_3 - u_1,$
$\alpha + \beta = 138°.2 = u_3 - u_1,$

Table 6.3 Observations of ADS 1227.

t	θ	ρ	No. of Nights	Observer	Instrument	Aperture (cm)
1908.87	267°.0	0".22	3	Aitken	Lick refractor	91
20.68	247.0	22	5	Aitken	Lick refractor	91
30.29	237.5	20	2	van Biesbroeck	Yerkes refractor	102
34.09	242.9	23	2	Aitken	Lick refractor	91
48.78	205.8	18	2	van Biesbroeck	McDonald reflector	210
58.66	182.3	20	3	van den Bos	McDonald reflector	210
61.73	181.5	18	2	Couteau	Yerkes refractor	102
65.051	173.6	19	4	Worley	Washington refractor	66
65.794	170.0	22	3	Worley	Flagstaff reflector	102
67.78	165.2	17	3	Couteau	Nice refractor	50
69.74	157.7	17	2	Couteau	Nice refractor	50
76.81	144.6	20	1	Couteau	Nice refractor	74

131 Computation of Orbits and Stellar Masses

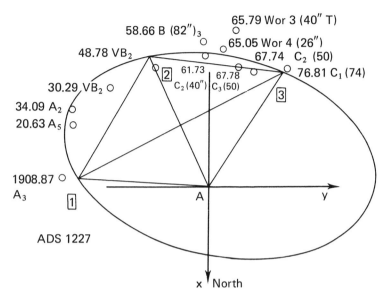

Figure 6.6 The Method of Thiele, Innes, and van den Bos. The orbit of the double star ADS 1227 = A 1913 is represented. The observations are marked, with their epochs, abbreviation for the observer's name, the number of nights, and the aperture of the instrument in inches (″) or centimeters. The apparent ellipse was drawn according to the three normal places chosen.

Table 6.4 Constant of Areas for ADS 1227.

t	Δt	ΔS	$c = \Delta S/\Delta t$	c
1908.87				
	11.81	0.212	0.0180	
20.68				0.0174
	28.10	483	172	
48.78				
	19.00	335	176	
67.78				
	9.03	164	182	
76.81				0.0178

from which we find

$e \sin u_2 = +0.20979,$
$e \cos u_2 = -0.24825,$

which gives $u_2 = 139°.8$, $e = 0.32$, then $u_1 = 50°.8$, $u_3 = 189°.0$. These three values of u can be used to calculate the mean anomalies

$M_1 = u_1 - e \sin u_1 = 0.63862$ radians,
$M_2 = u_2 - e \sin u_2 = 2.2342$ radians,
$M_3 = u_3 - e \sin u_3 = 3.34673$ radians,

which lead to the times of periastron passage

$T = t_1 - M_1/\mu = 1892.90,$
$T = t_2 - M_2/\mu = 1892.99,$
$T = t_3 - M_3/\mu = 1893.07,$

which can be rounded off to $T = 1893.0$ or $T + P = 2050.1$.

It remains to compute the constants A, B, F, and G. We have

$x_1 = -0''.01005,$ $y_1 = -0''.19174,$
$x_2 = -0''.18006,$ $y_2 = -0''.08705,$
$x_3 = -0''.15677,$ $y_3 = +0''.10734,$
$X_1 = +0.31203,$ $Y_1 = +0.73420,$
$X_2 = -1.08380,$ $Y_2 = +0.61151,$
$X_3 = -1.30769,$ $Y_3 = -0.14821.$

Solution of the equations

$x_i = AX_i + FY_i, \qquad y_i = BX_i + GY_i,$

for $i = 1,3$ and $i = 1,2$ gives

$A = +0''.1276,$	$A = +0''.1276,$
$B = -0''.0551,$	$B = -0''.0549,$
$F = -0''.0679,$	$F = -0''.0683,$
$G = -0''.2377,$	$G = -0''.2397,$

respectively. We adopt

$A = +0''.1276,$
$B = -0''.0550,$
$F = -0''.0681,$
$G = -0''.2387,$

which give the Campbell elements by equations (6.14). We have, then, the following set of elements:

$P = 157.1$ years,
$n = 2°.29153$,
$T = 2050.1$,
$a = 0\farcs25$,
$e = 0.32$,
$i = 123°.4$,
$\Omega = 77°.4$,
$\omega = 95°.9$,

$A = +0\farcs1276$,
$B = -0\farcs0550$,
$F = -0\farcs0681$,
$G = -0\farcs2387$.

Table 6.5 gives the differences between the observed and computed values, denoted by O − C and called "residuals." Also given again there are the dates, the observations, and the abbreviated names of the observers. Apart from the measure made in 1934, the observations are represented correctly. This computation is based on too short a part of the orbit to give certain elements. It is what is called a preliminary orbit. The ephemeris below will enable the accuracy of the result to be verified in the years to come.

1978.0	143°.1	0″.19
80.0	139.1	0.20
82.0	135.1	0.20
84.0	131.7	0.20

Danjon's Method of Opposite Points

This method is explained in detail in the *Bulletin Astronomique* (vol. 11, 1938) and in the *Annales de l'Observatoire de Strasbourg* (vol. 5, 1956). It is an analytical method; there is no need to draw the apparent ellipse, except in some cases to obtain an exact value for the period. The method is best applied to double stars that have described at least one orbital revolution. Quite certain values of the elements can be obtained in this way, since all the observations contribute to the computation, which, in turn, provides a strict check

134 Computation of Orbits and Stellar Masses

Table 6.5 Residuals for ADS 1227.

t	θ	ρ	No. of Nights	Observer	O − C	
1908.87	267°.0	0″.22	3	A	−0°.2	+0″.03
20.68	247.0	22	5	A	−1.2	− 01
30.29	237.5	20	2	VBs	+2.4	− 03
34.09	242.9	23	2	A	+13.2	00
48.78	205.8	18	2	VBs	−0.1	− 02
58.66	182.3	20	3	B	−3.5	+ 02
61.73	181.5	18	2	C	+2.5	00
65.051	173.6	19	4	Wor	+2.2	+ 01
65.794	170.0	22	3	Wor	+0.2	+ 04
67.78	165.2	17	3	C	0.0	− 01
69.74	157.7	17	2	C	−3.1	− 01
76.81	144.6	20	1	C	−0.9	+ 01

on the measures. The method can be applied several successive times to the observations, thus converging to the final result. The computer, protected from surprises, is certain to arrive eventually at a result very close to the truth, as far as the measures permit. Thus he saves himself the sometimes long and risky process of differential corrections to the elements. The method of opposite points is recent. It is the one most used, after that of Thiele, Innes, and van den Bos, which, like all methods based on drawing the apparent trajectory, requires extreme care.

"Opposite points" is the name for pairs of observations whose position angles differ by 180°. Their true anomalies, therefore, must also differ by 180°. This property is independent of the inclination, like the law of areas, but it is specific to Keplerian motion, which the law of areas is not. The method comprises two parts: first the computation of the dynamic elements, then that of the Thiele-Innes constants.

For each pair of opposite points, Thiele's fundamental equation, (6.28), can be simplified, since the area S_2 is zero. We have

$$M_2 - M_1 = u_2 - u_1 - \sin(u_2 - u_1). \tag{6.34}$$

Now $v_2 - v_1 = 180°$, $\theta_2 - \theta_1 = 180°$, but neither $M_2 - M_1$ nor $u_2 - u_1$ is 180°. Let us write

$$M_2 - M_1 = 180° - \Delta,$$

where

$$\Delta = 180° - n(t_2 - t_1). \tag{6.35}$$

Danjon called Δ the *characteristic angle*. It can also be written

$$\Delta = (v_2 - M_2) - (v_1 - M_1). \tag{6.36}$$

We see that Δ is the algebraic difference of the equations of the center for the two opposite points. It is zero at periastron and apastron, negative between apastron and periastron, and positive elsewhere. We will show that Δ is a function of M_1 (or M_2) and the eccentricity. We have

$$\tan(v/2) = \left(\frac{1+e}{1-e}\right)^{1/2} \tan(u/2),$$

which implies, for two opposite points,

$$\tan(u_2/2) = -\frac{1-e}{1+e} \cot(u_1/2);$$

then

$$\tan\frac{u_2 - u_1}{2} = -\frac{1 - e\cos u_1}{e\sin u_1} = \frac{-(1-e^2)^{1/2}}{e\sin v_1}.$$

Since v_1 is a function of M_1 (or M_2) and e, this shows that Δ is also a function of M_1 (or M_2) and e.

Bivariate tables giving M as a function of Δ and e have been computed by Muller. These tables, published in the *Annales de l'Observatoire de Strasbourg*, are seldom available outside observatories. We give an extract from them in table 6.6. With its help, thanks to the judicious choice of tabular intervals, the method can be applied at the cost of some easy graphical interpolations. Danjon called the curves of $\Delta(M)$ for constant e *characteristic curves*; he remarked that the maximum value of Δ is attained when $v_1 = 90°$. Thus,

$$M_1 = \cos^{-1} e - e(1-e^2)^{1/2}$$

and

$$\Delta_{max} = 180° - 2M_1.$$

Table 6.6 Extract from Muller's Tables.

Δ	e=0.1	e=0.2	Δ	e=0.3	e=0.4	Δ	e=0.5	e=0.6	Δ	e=0.7	e=0.8	e=0.9
0	0.00	0.00	0	0.00	0.00	0	0.00	0.00	0	0.00	0.00	0.00
1	2.03	0.80	3	1.23	0.67	4	0.50	0.27	10	0.32	0.13	0.03
2	4.06	1.60	6	2.46	1.35	8	1.00	0.53	20	0.65	0.26	0.06
3	6.10	2.40	9	3.69	2.03	12	1.51	0.80	30	0.98	0.38	0.08
4	8.15	3.21	12	4.94	2.71	16	2.02	1.07	40	1.33	0.52	0.11
5	10.22	4.01	15	6.21	3.40	20	2.53	1.35	50	1.69	0.65	0.15
6	12.32	4.82	18	7.50	4.10	24	3.05	1.62	60	2.07	0.80	0.18
7	14.45	5.63	21	8.81	4.81	28	3.58	1.90	70	2.49	0.95	0.21
8	16.62	6.45	24	10.16	5.53	32	4.12	2.19	80	2.96	1.12	0.25
9	18.84	7.27	27	11.55	6.26	36	4.68	2.47	90	3.49	1.31	0.29
10	21.11	8.09	30	12.98	7.00	40	5.25	2.77	100	4.09	1.52	0.33
11	23.44	8.93	33	14.47	7.64	44	5.84	3.07	105	4.45	1.64	0.36
12	25.85	9.76	36	16.04	8.55	48	6.44	3.38	110	4.85	1.77	0.38
13	28.35	10.61	39	17.67	9.35	52	7.08	3.71	115	5.30	1.92	0.41
14	30.96	11.46	42	19.40	10.19	56	7.74	4.04	120	5.82	2.08	0.44
15	33.70	12.33	45	21.26	11.05	60	8.44	4.38	125	6.46	2.26	0.47
16	36.61	13.20	48	23.29	11.95	64	9.17	4.74	130	7.24	2.47	0.51
17	39.73	14.08	51	25.48	12.89	68	9.95	5.11	132	7.63	2.56	0.52
18	43.11	14.98	54	27.96	13.87	72	10.79	5.50	134	8.05	2.66	0.55
19	46.86	15.88	56	29.80	14.57	76	11.70	5.91	136	8.56	2.77	0.57
20	51.14	16.81	58	31.84	15.28	80	12.69	6.35	138	9.15	2.89	0.59
21	56.28	17.74	60	34.17	16.03	84	13.80	6.82	140	9.88	3.02	0.60

137 Computation of Orbits and Stellar Masses

22	18.70	61	35.46	16.41	88	15.03	7.32	142	10.84	3.16	0.62
23	19.67	62	36.89	16.81	92	16.46	7.87	143	11.46	3.24	0.64
24	20.66	63	38.48	17.22	96	18.16	8.47	144	12.24	3.32	0.65
25	21.68	64	40.30	17.63	100	20.29	9.14	145	13.34	3.41	0.66
26	22.71	65	42.40	18.06	102	21.60	9.50	146	15.62	3.50	0.68
27	23.78	66	45.04	18.50	104	23.18	9.89	148		3.70	0.71
28	24.87	67	48.87	18.95	105	24.12	10.09	150		3.94	0.73
29	26.00	68		19.42	106	25.21	10.30	152		4.23	0.77
30	27.16	70		20.40	107	26.51	10.52	154		4.58	0.81
31	28.36	72		21.46	108	28.25	10.75	155		4.79	0.83
32	29.61	74		22.60	109	30.68	10.99	156		5.04	0.85
33	30.90	76		23.86	110		11.23	157		5.33	0.88
34	32.26	78		25.25	112		11.76	158		5.69	0.90
35	33.68	80		26.82	114		12.35	159		6.14	0.93
36	35.17	81		27.69	116		13.00	160		6.80	0.96
37	36.76	82		28.64	118		13.74	161		8.06	0.99
38	38.45	83		29.67	120		14.60	163			1.07
39	40.28	84		30.82	122		15.64	165			1.16
40	42.27	85		32.13	123		16.23	167			1.29
41	44.47	86		33.64	124		16.92	169			1.46
42	46.97	87		35.49	125		17.71	170			1.60
43	49.91	88		37.95	126		18.67	171			1.78
44	53.58	89		42.47	127		19.90	172			2.05
45	59.04	90			128		21.71	173			2.64
	63.18										

138 Computation of Orbits and Stellar Masses

The table gives the mean anomaly in degrees for each value of e and Δ. For each value of the eccentricity, it stops at a value close to the maximum value of Δ. The values for points on the other side of the maximum are deduced by symmetry:

$M'_1 = 180° - M_1 - \Delta$.

A few remarks will clarify the matter. Suppose that we wish to compute the elements of a binary that has been observed for at least one revolution. We determine its period and then draw the curve $\theta(t)$ after having reduced all the observations to one revolution (fig. 6.7). This curve should be drawn to a large scale; Danjon recommended several centimeters per year and several millimeters per

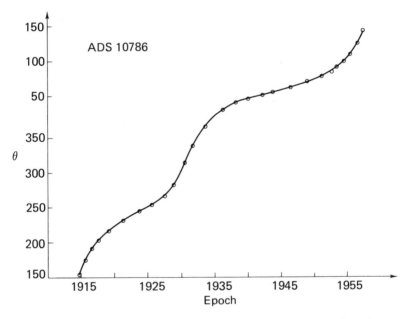

Figure 6.7 Graph of $\theta(t)$ for the companion star with respect to the primary in the system ADS 10786. The circles represent some of the observations between 1915 and 1960. The original working graph had a scale of 5 mm per degree and 5 cm per year, and it required a piece of paper 2 m by 1.80 m. The parts where the slope is steepest correspond to the times when the separation is smallest.

degree, so that each individual observation could be plotted. The results of the computation will enable corrections to be made to the curve, if necessary.

Mark each pair of opposite points with the suffixes 1 and 2, and make a table of a series of pairs, for example (0°, 180°), (10°, 190°), ..., (170°, 350°). Write down their epochs, t_1, t_2, ...; then $m_1 = nt_1$, $m_2 = nt_2$, and finally Δ. The maximum value of Δ gives an approximate value for e. Use this to determine M_1 with the help of the tables, and thus obtain T. In fact,

$$m_1 - M_1 = nt_1 - n(t_1 - T) = nT.$$

The differences $m_1 - M_1$ (or $m_2 - M_2$) ought to be constant for all pairs of points. If they are not, a slightly different eccentricity should be chosen by interpolation, until the table of the differences $m_1 - M_1$ does not show any systematic deviation. This will provide values of e and T at the same time. Having thus obtained those two elements, calculate

$$M_1 = m_1 - nT$$

and

$$M_2 = m_2 - nT,$$

from which v_1 and v_2 can be obtained from tables of Keplerian motion. The differences $v_2 - v_1$ ought to be close to 180°; if they are not, the drawing of the curve $\theta(t)$ must be at fault. The values thus obtained for the true anomalies are called *observed true anomalies*, and are denoted by v_O. Now we are going to compare them with the true anomalies computed by the equations of condition. For these, Danjon derived from the relations (6.11) and (6.12)

$$a\rho \cos\theta = r(A\cos v + F\sin v)$$

and

$$a\rho \sin\theta = r(B\cos v + G\sin v);$$

then he eliminated a, ρ, and r to find

$$\tan\theta = \frac{B + G\tan v}{A + F\tan v}. \tag{6.37}$$

By using the pairs (v_0, θ), one can write as many equations of this type as there are points chosen, but they only allow us to determine three of the constants as a function of the other. Danjon set $A = \alpha F$, $B = \beta F$, and $G = \gamma F$, so the equation became

$$\tan\theta = \frac{\beta + \gamma \tan v}{\alpha + \tan v}. \tag{6.38}$$

He then found α, β, and γ by the method of least squares. It was shown by P. J. Morel, however, that the choice of the divisor in equation (6.37) is not unimportant. It is preferable to choose the largest constant. Since the constants are unknown at the beginning, the first choice is arbitrary. It is especially desirable to avoid one that proves to be very much smaller than the others; if necessary, stop and begin the solution of equation (6.38) again after the first attempt.

The constants α, β, and γ provide, in their turn, calculated true anomalies by the equations (6.38). From these anomalies v_c, calculate mean anomalies M_c. Determine the deviations $v_0 - v_c$ and then $M_c - M = \delta M$, which, converted into time, give

$$\delta t = \frac{1}{n} \delta M.$$

These are the corrections to be made to the curve $\theta(t)$.

It may be that the values of δt are not significant and do not change the course of the curve. If they prove to be very large, they will have to be taken into account by drawing a new curve $\theta(t)$ as a second approximation, and repeating the whole computation from the beginning. Usually, this second approximation is amply sufficient.

Return to the equations (6.38) and, with the help of the constants derived, calculate the position angles at the epoch of each observation. The run of the residuals shows the quality of the elements. It remains to determine the last constant. We proceed thus. The three Thiele-Innes constants, as functions of the fourth, enable the geometric elements i, ω, and Ω to be computed by the first three of the relations (6.14). Then compute the separations by supposing the major semiaxis to be equal to unity, and plot the graph of ρ_c/a

141 Computation of Orbits and Stellar Masses

against ρ_0. The representative points should be distributed close to a straight line of slope a. Thus, all the elements are known. We note that just one observation of separation is sufficient to determine the major semiaxis. Thus, the great advantage of Danjon's method is that the elements are determined only from the position angles (which are much better measured than the separations, which serve only to give the scale of the orbit). The best of the observed separations are sufficient for this purpose.

Example

The orbit of the double star ADS 10786 = AC 7 mag 10.5–11.0 α(1950) $7^h44^m\!.5$, δ(1950) $+27°45'$ was computed by the author in 1959. The elements represent later observations well. We will reproduce their run at the end of the computation. This pair of red dwarfs has completed its third revolution since its discovery by Alvan Clark in 1856. Although very faint, it has been followed regularly, and the first orbit was calculated in 1878. The method of opposite points is particularly suitable for this star, since its faintness makes the measurement of separation difficult and imprecise, while the relatively large distance between the components allows fairly certain measures of position angles. Selected measures are given at the end of the calculation.

After having determined the period as 43.20 years, we have reduced all the observations to the equinox of 1950 and to one and the same revolution, then drawn the curve $\theta(t)$ to a scale of 5 mm per degree and 5 cm per year. Next, for 36 position angles at 10° intervals, we have computed the characteristic angle.

$$\Delta = (v_2 - M_2) - (v_1 - M_1) = 180° - n(t_2 - t_1),$$

where t_1 and t_2 are the epochs of passage through two opposite points. The zero of time is 1889.0. A table is made giving the position angle every 10° from 0° to 180°, the anomalies $nt_1 = m_1$ and $nt_2 = m_2$, and the characteristic angle. This column and that of t_2 show that periastron is between epochs 31.68 and 33.43. On the other hand, the maximum of Δ shows that the eccentricity is close to 0.18. Obtain, then, the values of M for $e = 0.18$ and 0.19 from Muller's tables, then the differences $m_2 - M = m_0$. They are clearly systematic for $e = 0.19$. Values of M and m_0 corresponding to the neighborhood of the maximum value of Δ are omitted because they are less accurate.

Take a provisional mean value of $276°.3$ for m_0, and choose $e = 0.18$. Calculate the values $M_{1,2} = m_{1,2} - 276.3$ for the 36 values of θ. These are the observed mean anomalies. The observed true anomalies are deduced from them with the help of tables of Keplerian motion. Note that the differences $v_2 - v_1 - 180°$, theoretically zero, are small; two-thirds of them are less than a degree. With the help of the largest residuals, the corresponding anomalies $M_{1,2}$ are corrected, which give, in their turn, an improved value of m_0, which is $276°.66$ (whence $T = 1922.20$), and of e, which is adjusted to 0.178. The table for Keplerian motion then gives the observed true anomaly v_0, which is to be found in table 6.8.

This done, the 36 equations (6.38) can be written, with the values of θ and v_0 taken from the table. As the residuals are small, add the equations three by three before solving them by the method of least squares. They could even be added in groups of 12; this would give an almost identical result more quickly. The equations give the following values for the constants:

$$\alpha = -1.749, \quad \beta = -2.831, \quad \gamma = -0.957.$$

The true anomalies v_c calculated from these constants are included in table 6.8, together with the differences $v_0 - v_c$. These differences are small, and correspond to minor corrections of the order of a few hundredths of a year to the epochs t_1 and t_2. Check that the corrections thus introduced into the curve $\theta(t)$ do not modify it appreciably (except, perhaps, around the position angle 250°), but the curve tends to approximate the observations better.

The geometric elements are then obtained by the classical formulae:

$$i = 66°.2, \quad \Omega = 60°.7, \quad \omega = 174°.0.$$

It remains to determine the major axis. Since the measures are fairly scattered, and the observations are made with instruments of very diverse apertures, it is necessary to see whether there are any systematic deviations from one observer to the other, or for one individual as a function of the separation.

143 Computation of Orbits and Stellar Masses

Table 6.7 Computation of the Orbit of ADS 10786.

θ	t_1	t_2	m_1	m_2	Δ	e = 0.18		e = 0.19	
						M	m_o	M	m_o
0°	0°.64	26°.88	5°.3	224°.0	−38°.7	−51°.8	275°.8	−44°.8	268°.8
10	1.43	27.44	11.9	228.7	−36.8	−46.0	274.7	−40.6	269.3
20	2.41	28.15	20.1	234.6	−34.5	−40.7	275.3	−36.4	271.0
30	3.67	29.07	30.6	242.2	−31.6	−35.3	277.5	−32.0	274.2
40	5.53	30.20	46.1	251.7	−25.6	−26.6	278.3	−24.8	276.9
50	8.54	31.68	71.2	264.0	−12.8	−12.2	276.2	−11.3	275.3
60	12.43	33.43	103.6	278.7	+4.9	+4.6	274.1	+4.9	273.0
70	16.18	35.54	134.8	296.2	18.6	18.3	277.9	16.8	279.4
80	19.12	37.22	159.3	310.2	29.1	31.4	278.8	27.5	283.5
90	21.07	38.50	175.6	320.8	34.8	41.3	279.5	36.3	284.9
100	22.39	39.39	186.6	328.2	38.4	50.8	277.4		
110	23.28	40.14	194.0	334.5	39.5	55.2	279.3		
120	24.00	40.79	200.0	339.9	40.1				
130	24.58	41.24	204.8	343.7	41.1				
140	25.09	41.72	209.1	347.7	41.4				
150	25.49	42.23	212.4	351.9	40.5				
160	25.88	42.73	215.7	356.1	39.6	84.8	271.3		
170	26.35	0.07	219.6	0.6	39.0	88.1	272.5		

Table 6.8 Computation of the Orbit of ADS 10786.

θ	v_o	v_c	$v_o - v_c$	θ	v_o	v_c	$v_o - v_c$
0°	−71°.4	−71°.3	−0°.1	180°	108°.6	108°.7	−0°.1
10	−65.7	−65.8	+0.1	190	114.6	114.2	+0.4
20	−58.4	−58.9	+0.5	200	121.8	121.1	+0.7
30	−48.6	−49.9	+1.3	210	130.7	130.1	+0.6
40	−36.0	−37.2	+1.2	220	142.9	142.8	+0.1
50	−18.6	−19.2	+0.6	230	161.6	161.4	+0.2
60	+2.7	+4.2	−1.5	240	−175.2	−175.8	+0.6
70	27.8	28.1	−0.3	250	−152.6	−151.9	−0.7
80	47.0	46.9	+0.1	260	−133.6	−133.1	−0.5
90	60.5	60.2	+0.3	270	−120.1	−119.8	−0.3
100	69.5	69.7	−0.2	280	−110.2	−110.3	+0.1
110	76.9	76.8	+0.1	290	−103.3	−103.2	−0.1
120	82.9	82.5	+0.4	300	−97.4	−97.5	+0.1
130	87.0	87.3	−0.3	310	−92.5	−92.7	+0.2
140	91.2	91.6	−0.4	320	−88.0	−88.4	+0.4
150	95.5	95.6	−0.1	330	−84.5	−84.4	+0.1
160	99.7	99.7	0.0	340	−80.8	−80.3	−0.5
170	104.0	104.0	0.0	350	−76.5	−76.0	−0.5

From all the observations, the graph of ρ_c/a versus ρ_o has been drawn. The measures fall near a straight line that cuts the ρ_c/a axis just below the origin. In contrast with the usual situation for bright pairs, the observers have measured the components too close when the separation is small—which might not have been noticed had the elements been computed from the apparent ellipse. Finally, by basing our computation on measures made at epochs of wide separation, we find that the major semiaxis $a = 1''.360$, which enables the constants A, B, F, and G to be computed. Note that F, chosen as the divisor, is not the smallest of the four.

Elements

$P = 43.20$ years	$a = 1''.360$	$A = -0''.7118$
$n = 8°.3333$	$i = 66°.2$	$B = -1.1522$
$T = 1965.40$	$\Omega = 60°.7$	$F = +0.4070$
$e = 0.178$	$\omega = 174°.0$	$G = -0.3895$

145 Computation of Orbits and Stellar Masses

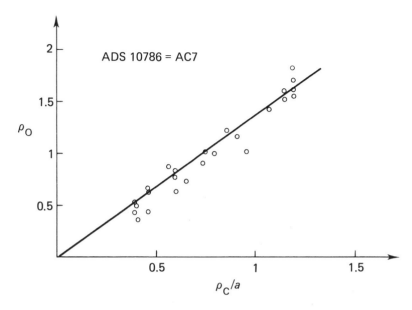

Figure 6.8 Determination of the major axis for ADS 10786. The points cluster around a straight line whose slope gives the major semiaxis. Only some of the observations are shown.

Ephemeris

1976.5	6°.6	0″.68	1980.5	38°.2	1″.15
7.5	18.1	0.79	1.5	42.4	1.26
8.5	26.8	0.92	2.5	46.0	1.36
9.5	33.2	1.04	3.5	49.1	1.44

Table 6.9 gives a series of observations and their residuals from the orbit. In order not to overload the text, we have not put in all the measures. We indicate only the number of nights in the cases where the means group the results of more than two observers.

Differential Corrections for the Elements

A properly done computation ought to provide acceptable residuals immediately. If it does not, it is useless to hope for a completely satisfactory improvement. The observations may be systematically

146 Computation of Orbits and Stellar Masses

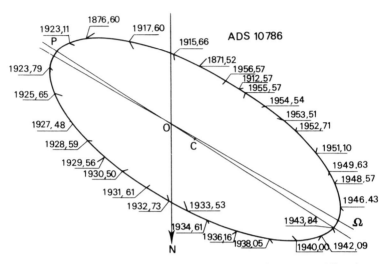

Figure 6.9 The apparent orbit of ADS 10786 = AC7. Only some of the observations are shown. The primary star is at O. The maximum elongation of the companion, in 1944 and 1987, is 1."65. Remember that an arc-second is the angle subtended by 1 cm seen from a distance of 2 km. The strokes show the deviation of the observed positions from the calculated ones.

in error, or an unobservable third body may be perturbing the motion of the pair. The importance and appearance of the deviations depend on how difficult the binary is to measure. A 10° error for a close pair at 0."2 matters less than a 2° deviation for Castor or γ Virginis. This is one of the reasons why the computation of orbits ought to be undertaken only by regular observers who know the difficulties of measurement well.

It is the deviations in θ that are significant and can, under certain conditions, be treated as differentials; the deviations in separation can be systematic without being significant, and can seldom be treated as differentials. We have

$$\theta = f(\Omega, i, \omega, e, T, n).$$

The total differential can be written

$$d\theta = \sum_{i=1}^{6} \frac{\partial f}{\partial x_i} dx_i, \tag{6.39}$$

Computation of Orbits and Stellar Masses

Table 6.9 Selected Observations of ADS 10786.

Epoch	Observations		No. of nights	Observer[a]	Residuals	
1857.50	59.°2	1."82	2	Dawes	+0.°4	+0."22
65.79	84.1	1.17	7	Dembowski (5), Struve (2)	−0.2	+ .06
78.08	233.5	1.03	13		−0.2	− .05
85.56	286.7	0.67	5	H. Struve (2), Hall (3)	+1.8	.00
94.74	41.4	1.23	30		+0.3	+ .01
1902.94	63.6	1.63	32		−0.3	+ .05
06.28	73.0	1.37	18		0.0	− .02
08.46	80.1	1.22	22		−1.0	+ .05
11.59	102.4	0.72	8		+0.3	− .05
15.66	175.4	0.49	7		−0.5	− .06
21.11	233.0	1.00	14		+0.4	− .07
28.59	283.3	0.66	3	Aitken	+1.0	− .03
33.53	8.7	0.64	7		−0.6	− .06
40.00	47.6	1.37	36		−0.3	− .05
46.43	63.7	1.60	26		−0.6	+ .03
51.10	79.5	1.26	8	Baize 4, Baize 4	+1.1	+ .03
54.54	99.9	0.77	7	Couteau 3, Baize 4	+0.4	− .03
56.57	126.8	0.55	9	Worley 4, Baize 5	+0.1	− .03
60.65	204.0	0.70	4	Worley	+2.8	+ .10
64.68	234.2	1.07	4	Baize	0.0	− .01
74.54	330.8	0.51	7		−1.2	− .03

a. Numbers indicate number of measurements.

where the x_i are each one of the six elements. Therefore,

$\theta_0 - \theta_c = Ad\Omega + Bdi + Cd\omega + Dde + EdT + Fdn.$

From the relations (6.4) and (6.9), $A, \ldots, F$ can easily be computed. We find

$A = 1,$
$B = -\sin i \tan(v + \omega) \cos^2(\theta + \Omega),$
$C = \cos^2(\theta - \Omega) \sec^2(v + \omega) \cos i,$

$D = C \dfrac{2 - e\cos u - e^2}{(1 - e\cos u)^2} \sin u,$

$E = C \dfrac{1 - e^2}{(1 - e\cos u)^2},$

$F = E(t - T).$

Each observation gives an equation of the form (6.39). The whole set is solved by the method of least squares. This is a cumbersome process and, in fact, is little used. It is quicker to correct the elements by practical considerations based on the run of the residuals. Thus, a change in Ω changes all the position angles in the same way. It is convenient to write

$dM = (t - T)dn - ndT.$

The values of n and T can be adjusted to improve the residuals in two regions. If the eccentricity is large, a change in T will be more important than if it is small.

COMPUTING THE MASSES OF STARS

By Kepler's third law, the mass of a system can be obtained directly from

$$M_A + M_B = \dfrac{a''^3}{p''^3 P^2}, \qquad (6.40)$$

where M_A and M_B are the masses of the components, that of the sun being taken as unit; p'' is the trigonometrical parallax in arc-seconds; a'' is the apparent major semiaxis in the same units; and P is the period in years. This is the way in which fundamental masses, determined solely by measures of position, are obtained. Progress is limited by our knowledge of parallaxes, which are so small that we cannot envisage any improvement in their measurement unless there is a profound revolution in our techniques. We recall that the parallax is the angle subtended at the star by the astronomical unit or mean distance from the earth to the sun. By contrast, the major axes and the periods become better known with time, as observations are accumulated. Thus, of 750 known orbits, fewer than 50 have a parallax well enough measured to give good values for the masses. These results indicate that stars have masses in the range from 0.07 to 20 times the mass of the sun. Nine-tenths of the stars have masses between 0.4 and 2. The sun is in the middle of the scale.

The relation (6.40) gives the total mass of the system, but not the individual masses. To know them, we must know the motions of the components about the center of gravity of the system. This is no longer a task for visual observation, but for spectrographic or photographic observation. For the sake of completeness, however, we cannot leave a problem as important as this in the dark; we must study its essentials.

Mass Ratios from Radial Velocities

The masses of a pair can be determined by measuring the radial velocities of each component at several epochs. We will follow van den Bos's terminology (*Astronomical Techniques,* 1963). Let V_A and V_B be the radial velocities of the components A and B, and let V_G be that of the center of masses. Let us write

$V_R = V_B - V_A,$
$\beta = M_B/M_A,$
$kp = \beta(1 + \beta),$
$\ell p = 1/(1 + \beta),$

where p is the parallax and k and ℓ are constants. To fix our ideas, let us suppose V_R to be positive, which implies that the component B is receding. We then have

$$V_A = V_G - kpV_R, \qquad V_B = V_G + \ell p V_R. \tag{6.41}$$

Now, pV_R is known as a function of time and of the elements of the orbit from equations (6.15) and (6.16). The radial velocities are measured at different epochs, and a graph is drawn with pV_R on the abscissa and V_A and V_B on the ordinate. By equation (6.41), there should be two segments of straight lines that intersect when $pV_R = 0$ at a point whose ordinate is V_G. Thus the velocity of the center of mass is obtained. The slopes of the segments V_A and V_B are of opposite signs; their magnitudes give k and ℓ, and thus the mass ratio and the parallax, since

$$M_B/M_A = k/\ell,$$
$$p = 1/(k + \ell). \tag{6.42}$$

If V_A and V_B cannot be measured independently, but only V_R, the absolute parallax can be obtained, and the sum of the masses, but not the individual masses.

Computation of stellar masses from measurements of radial velocities is limited by the precision of spectrographs. The straight-line segments of figure 6.10 are as short as the velocity variation is small; their slopes are not always easy to obtain. Finally, the method can only be applied to well-separated binaries (at least 2 arc-seconds) and fairly bright ones (for example, not fainter than sixth magnitude). As a result, it has been applied to only a few pairs, such as γ Virginis, α Centauri, 70 Ophiuchi, and ξ Ursae Majoris. Nevertheless, modern techniques of electronic spectrography should allow increases in both the speed and the precision of radial-velocity measurement.

Mass Ratios from Photography

Photography of double stars with instruments of long focal length makes possible the determination of the positions of the components with respect to those of apparently neighboring stars. In some cases, sufficient plates have been accumulated to disentangle the complicated skein of the apparent movements of the components. The number of pairs that can be treated in this way is considerably greater than the number for which radial velocity measurements can be measured.

Generally, on one plate, the components of a double star are resolved if their separation is greater than 1".5. Below this separation,

151 Computation of Orbits and Stellar Masses

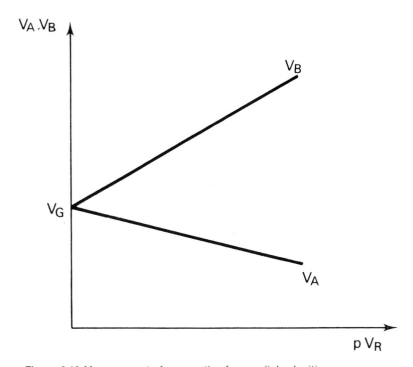

Figure 6.10 Measurement of mass ratios from radial velocities.

photographic images, much larger than the Airy disks, impinge upon each other. The maximum of photographic intensity, or the photocenter, corresponds neither to the position of the primary nor to that of the secondary. Let A and B be the positions of the components (fig. 6.11), G that of the center of mass, and P that of the photocenter. These four points lie in the same straight line. Some investigators (van de Kamp, for example) determine the position of the photocenter by the condition

$$\frac{AP}{AB} = \gamma = \frac{\ell_B}{\ell_A + \ell_B} = \frac{1}{1 + 10^{0.4\Delta m}}, \tag{6.43}$$

where ℓ_A and ℓ_B are the luminosities of the components and Δm is the difference in their magnitudes. Equation (6.43) is not rigorous. Various authors have discussed it; in particular, Morel (1969) has

152 Computation of Orbits and Stellar Masses

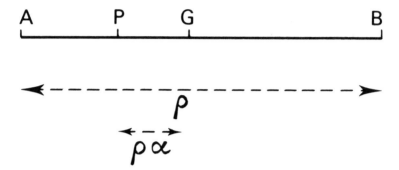

Figure 6.11 Measurement of mass ratios from position of photocenter.

worked out an approximate theory of the diffusion of light in a photographic plate. The problem is important because it is a keystone of our knowledge of stellar masses. The results of equation (6.43) and Morel's theory are given in table 6.10. They differ appreciably for large differences of brightness; but it should be remembered that large differences of brightness correspond to dissimilar stars, of different types, which do not affect the plate in the same way, and this modifies the function γ.

In the course of the orbital motion, the photocenter describes an orbit similar to that of the companion, with

$$\frac{PG}{AB} = \frac{\alpha}{A} \tag{6.44}$$

as the ratio of the axes. Now if, following van de Kamp, we write the mass ratio

$$R = \frac{AG}{AB} = \frac{M_B}{M_A + M_B}, \tag{6.45}$$

Table 6.10 Variation of γ with Δm.

	Δm						
	0	0.5	1	2	3	4	5
γ (van de Kamp)	0.5	0.387	0.285	0.137	0.060	0.025	0.01
γ (Morel)	0.5	0.386	0.284	0.125	0.05	0.01	0

we deduce that

$$R = \alpha/a + \gamma. \tag{6.46}$$

Measurement of the mass ratio of components not resolved photographically depends, therefore, on the measurement of α and on the assumptions made about the function γ. This granted, let us consider a plate of a double star. Two cases must be studied: when the components are resolved and when they are not.

In the first case, let us write the Cartesian coordinates X and Y of component A, oriented in the sense of increasing right ascension and declination:

$$\begin{aligned} X &= c_\alpha + \mu_\alpha(t - t_0) + pP_\alpha + a_\alpha RQ_\alpha, \\ Y &= c_\delta + \mu_\delta(t - t_0) + pP_\delta + a_\delta RQ_\delta, \end{aligned} \tag{6.47}$$

where c_α and c_δ are the heliocentric positions at the initial epoch t_0; μ_α and μ_δ are the annual proper motions projected on the axes; t is the epoch at which the plate was taken; P_α and P_δ are the parallactic factors, or projections of the magnitude of the parallax on the axes; a_α and a_δ are the projections of the major semiaxis, and Q_α and Q_δ are the projections of the orbital motion at the epoch t. The heliocentric positions, the proper motions, and the parallax are constants, but the parallactic factors and the orbital motion are not.

We refer the reader to specialized texts for the detailed study of parallactic factors. We simply give here the formulae by which they may be computed:

$$\begin{aligned} P_\alpha &= r(f\sin\odot + g\cos\odot), \\ P_\delta &= r(f'\sin\odot + g'\cos\odot), \end{aligned} \tag{6.48}$$

where r is the radius vector of the earth's orbit expressed in astronomical units, $\odot$ is the true longitude of the sun (both are given in almanacs), and

$f = +0.9174\cos\alpha,$
$g = -\sin\alpha,$
$f' = +0.3979\cos\delta - 0.9174\sin\alpha\sin\delta,$
$g' = -\cos\alpha\sin\delta.$

These constants contain the obliquity of the ecliptic, which varies slowly. The formulae are given for the year 2000. The factors Q_α and

154 Computation of Orbits and Stellar Masses

Q_δ are given by

$$Q_\alpha = \frac{1}{a}(AX + FY),$$

$$Q_\delta = \frac{1}{a}(BX + GY).$$

Note that X and Y in these equations are the reduced coordinates defined in equation (6.6)!

For a pair whose visual orbit is known, equations (6.47) contain four unknowns: the heliocentric positions, the components of proper motion, the parallax, and the mass ratio.

When the images of the components are not separated on the plate, only the motion of the photocenter can be followed. The last terms of each of the equations (6.47) become

$$\alpha Q_\alpha = (R - \gamma)Q_\alpha,$$
$$\alpha Q_\delta = (R - \gamma)Q_\delta. \tag{6.49}$$

In this case, the factor γ deprives the measurement of the mass ratio of its fundamental character. That is why the study of this still insufficiently known function is considered important.

Each observation provides an equation (6.47). But each plate must be reduced to the same conditions; otherwise the very small effects sought, which do not exceed 2–3 μm, will be submerged in the accidental errors. The causes of these errors are numerous, and include refraction, deviation of the plate from the focus, the plane of the plate not being perpendicular to the optical axis, and expansion caused by temperature changes. We do not try to eliminate these errors, but, rather, to reduce all the plates to the conditions of one of them in such a way that the results become comparable. Choose, in the neighborhood of the star to be studied, several fairly faint stars that can form a basic group to which the measures can be referred. Let $X_{c,i}$ and $Y_{c,i}$ be the coordinates of this basic group on the standard plate. All the plates must be reduced to this. Let x_i and y_i be the basic coordinates on any other plate; they are reduced to the standard plate by writing

$$X_{c,i} = A_0 + A_1 x_i + A_2 y_i,$$
$$Y_{c,i} = B_0 + B_1 x_i + B_2 y_i. \tag{6.50}$$

Since the field is small (a few arc-minutes), the equations can be limited to the first order; the effects of deformation are linear to a very high degree of approximation. The coefficients A and B are computed for each plate, by the method of least squares, and then they are substituted in the right-hand sides of equations (6.50) to obtain the coordinates X_i and Y_i reduced to the standard plate. The differences $X_{c,i} - X_i$ and $Y_{c,i} - Y_{c,i}$ should be negligibly small and random. If they are not, it is because at least one of the base group of stars has an appreciable proper motion or parallax, which vitiates the results. On the other hand, for the star being studied, we know the following from each plate:

$$X_0 = A_0 + A_1 x_0 + A_2 y_0,$$
$$Y_0 = B_0 + B_1 x_0 + B_2 y_0. \tag{6.51}$$

Thus the coordinates of the star under study are reduced to the standard plate. It is convenient to limit ourselves to three reference stars, provided we can be sure they are absolutely fixed.

Some astronomers, such as van de Kamp, use a method of reduction, called the *method of dependences,* in which weights are introduced as a function of the distance of the star being studied from the comparison star.

The coordinates provided by the equations (6.51), for each epoch of observation, serve to form the relations (6.47). The four unknowns are determined by solving these equations by the method of least squares. The solution provides three parameters: the proper motion, the parallax, and the mass ratio.

One such observational task is long; it can stretch over decades for each double star. Sproul Observatory and the Flagstaff station of the U.S. Naval Observatory specialize in these measures of masses and parallaxes. Under the leadership of van de Kamp, S. L. Lippincott, and K. Aa. Strand, they have accumulated tens of thousands of plates of nearby stars. Thousands of parallaxes have thus been calculated. The method can be used in the study of the orbits of binaries that are unresolved because they are too close or too faint. These studies are also aimed at the discovery of the first planets outside the solar system.

For the sake of completeness, it is useful to give some astrophysical information about spectral classification before presenting the

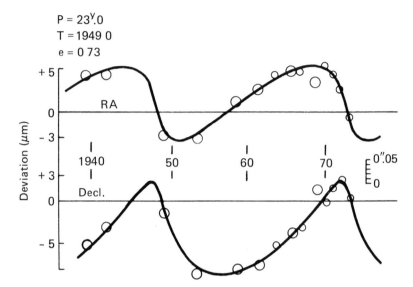

Figure 6.12 The perturbed trajectory of a star (BD + 67° 552) that has an invisible companion. Neither the motion in right ascension (RA) nor that in declination is rectilinear; they are looped because of the invisible companion. The deviations reach 5 μm at the focus of the Sproul refractor. The size of the circles represents the probable errors of observation. The photographic image itself is about 100 μm across. From S. L. Lippincott, *Astronomical Journal* 78 (1973): 303; reproduced by permission of the author and the editor.

table of the best-known masses. The stars differ not only in mass and volume, but also in temperature, chemical composition, and physical properties. Astronomers classify stars just as a botanist does plants. Classification is a sorting out that serves to clarify the laws, which in turn make it possible to explain the behavior of nature. The classification of stars has been the subject of weighty specialized treatises. As the reader may not be very familiar with this subject, we will give the bare essentials.

Stars have their own colors, which are related to their surface temperatures. A simple pair of binoculars shows this well. Vega and Sirius are white, Capella is yellow like the sun, Antares and Betelgeuse are clearly red. These different colors suggest a classification

157 Computation of Orbits and Stellar Masses

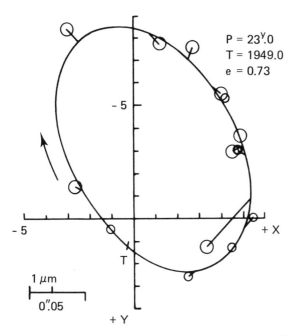

Figure 6.13 Orbit of the photocenter of the star BD +67° 552. This orbit is obtained by analysis of the trajectory shown in figure 6.12. The trajectory does not deviate by as much as 10 μm from a straight line, as seen at the focus of the Sproul refractor. The mass of the perturbing body is about one-third that of the sun. It is a small invisible star. From S. L. Lippincott, *Astronomical Journal* 78 (1973): 303; reproduced by permission of the author and the editor.

that Father Secchi thought of establishing in 1868 by distributing the stars into four classes, from the hottest to the coolest. Later, in 1890, Pickering and Flemming, as a result of precise spectroscopic observations, used the first capital letters to classify the stars according as they are hotter or cooler. At the beginning of this century, Cannon, at Harvard Observatory, laid down the broad lines of the classification still in use today. Beginning with the hottest stars, a sequence O,B,A,F,G,K,M can be formed. Each letter specifies a type, itself divided into ten subtypes. The sun is a G2 star; Sirius is an A1 star. Precise measures of surface temperature can be associated

158 Computation of Orbits and Stellar Masses

with these types, which range from 20,000°K for 80 to 3,000°K for M4.

Objects of different volumes may have the same temperature. For example, the sun and Capella are of the same type, but Capella is a system of two nearly identical bodies, each a thousand times as great as the sun, although their total mass does not exceed five times that of our great luminary. These are giant stars, characterized by low surface gravity and low pressures. Thus, a new parameter must be introduced: the pressure, which determines the size of a star. This new parameter defines what is called the *luminosity class.* It is denoted by Roman numerals, I designating the most luminous stars and VI the subdwarfs. The sun belongs to class V, the most numerous class and called the *main sequence.* Its complete classification, therefore, is G2V. Capella is formed of two giant stars, G5III and G0III. All stars fit into this classification, except the compact stars at the end of their evolution, such as white dwarfs and neutron stars (or pulsars). These stars, however, are difficult to observe; only a few are known.

Table 6.11 gives a list of 48 pairs for which trustworthy values of the total mass are known. The first line of each entry gives the number in Aitken's Catalogue (ADS) or a reference to another catalogue; the name of the object as a double star; its denomination in the constellation; magnitudes taken from the catalogue by Finsen and Worley (1970); spectral types and luminosity classes taken from Glese's catalogue (1969); the trigonometric parallax taken from the same catalogue, unless otherwise indicated; the mass of the system in solar masses, calculated by Kepler's third law; and the mass ratio $M_B/(M_A + M_B)$ taken from "Empirical Data on Stellar Masses, Luminosities, and Radii," by Harris, Strand, and Worley (*Basic Astronomical Data,* 1963), unless otherwise indicated. the second line gives celestial coordinates for 1950 in hours, minutes and tenths of a minute, then in degrees and arc-minutes; the computer of the orbit and and the year of computation, from the card index at Nice Observatory; the period in years; and the apparent major semiaxis in arc-seconds. N indicates that there is a note at the end of the table.

It is instructive to plot a graph of the masses and absolute magnitudes as shown in figure 6.14. Note that the masses of a system are at the ends of a straight line whose slope, in general, is very

Table 6.11 Well-determined Masses.

ADS	Name α 1950 δ	Name	Computer and Year	Magnitude	Spectrum	P(Years)	p_{tri} / a''	$\Sigma M_\odot$	R
48 AB	OΣ	547	Guntzel-Lingner (1954)	8.9-9.0	dK6-dM0	362.3	0".94	2.16	0.5
00028	N	4532		6.4-7.2	dG4-dG8		6.179	2.82	N*
61	Σ	3062					.045		0.59
00035	N	5809	Baize (1957)	6.3-6.4	G5V	106.83	1.432	1.40	N
520	β	395		3.5-7.5	G0IV-M0V	25.00	.70		0.5
00347	S	2502	van den Bos (1936)	7.8-7.9	K3V		0.670	1.42	0.39
671	Σ	60	η Cas			480	.174		N
00461	N	5733	Strand (1969)				11.9939	1.29	0.5
−30°529	δ	31 AB				4.5587	.057		
01327	S	3010	Wieth-Knudsen (1954)	12.5-12.9	M5.5-M5.5		0.1708	0.22	0.45
	L	726-8	Worley and Behall (1973)				.384		N
01364	S	1813		9.3-9.5	M1.5	26.52	2.06	0.86	0.54
1865	A	2329					.066		N
02251	N	0412	van den Bos (1962)	5.6-5.6	F8V	25.25	0.5405	0.54	0.5
−12°501	ϕ	312	ε Cet	3.9-6.5	F8IV	2.667	.069	2.03	–
02371	S	1205	Finsen (1970)				0.108		
2402	h	3555	α For	0.6-1.1	G5III-G0III	154.5	.74	4.50	–
03099	S	2911	van den Bos (1956)				2.700		
3841			Capella				.075		
05130	N	4557	Merrill (1921)	8.4-8.9	K5V	0.2848	0.05360	1.36	–
31°2902	Hu	1399 AB					0.49		

160 Computation of Orbits and Stellar Masses

Table 6.11 (cont.)

ADS	α 1950 δ	Name	Computer and Year	Magnitude	Spectrum	P(Years)	ρ_tri / a''	$\Sigma M_\odot$	R
05585	S	3102	Baize (1951)			72.0	0.94		
	Ross	614	Lippincott (1955)	10.9–14.4	M7V	16.6	14.248	0.18	0.351
06283	S	0246	Sirius	1.5–8.5	A1V-w.d.		0.904		N
5423	AGC	1	van den Bos (1960)			50.09	.377	3.14	0.33
06429	S	1639	Procyon	0.4–10.3	F5IV-w.d.		7.500		0.27
6251	Shaeb						.285		
07368	N	0522	Strand (1949)	5.6–6.2	GIV	40.65	4.548	2.46	0.53
6420	β	101	9 Pup			23.26	.062	1.81	N
07495	S	1346	Douglass (unpublished)	10.8–11.0	M1V		0.616	0.69	0.5
7114	Hu	628 BC					.066		
08558	N	4814	Eggen (1962)	4.1–6.2	F5V	39.69	0.68	2.10	0.427
+42°1956	Kui	37	10 UMa	7.2–7.2	KOV	21.81	0.064	1.08	N
08576	N	4200	Baize (1967)	7.9–8.0	K3V	2.65	0.640	1.26	0.5
+15°2003		ϕ347	81 Cnc				.064		0.51
09096	N	1512	Finsen (1966)	4.1–4.6	F2IV	34.20	0.1258	2.44	0.49
7284	Σ	3121							
09149	N	2847	van den Bos (1937)	3.5–3.5	F0V–F0V	34.11	0.660	1.84	0.49
−39°3651			ψ Vel				.065		
09287	S	4015	van den Bos (1945)				0.920		
8630	Σ	1670	γ Vir			171.37	.099		
12391	S	0111	Strand (1935)				3.746		

161 Computation of Orbits and Stellar Masses

ADS	Disc	Number	Author	Mag	Spectrum	Period	a	M	m
8804	Σ	1728	α Com	5.1–5.1	F5V	25.87	.053	2.91	0.51
13076	N	1747	Haffner (1938)				0.662	0.73	0.412
8862	Hu	644		9.2–10.0	dM2	49.18	.119		N
13176	N	4802	Baize (1969)				1.44	0.93	0.49
9031	Σ	1785		7.6–8.0	K6V–K6V		.086		
13468	N	2714	Strand (1953)			155.0	2.423	2.07	0.46
−60°5483			α Cen	0–1.2	G2V–K0V	79.92	.743		
14362	S	6038	Heintz (1959)				17.583	1.59	0.46
9413	Σ	1888	ξ Boo	4.7–6.9	G8V–K4V		.148		N
14491	N	1919	Wielen (1962)			151.505	4.9044	1.58	0.497
9617	Σ	1937	η CrB	5.6–5.9	G2V–G2V		.060		N
15211	N	3028	Danjon (1938)			41.56	0.839	2.25	0.478
9716	OΣ	298		7.5–7.6	K4V		.041		
15343	N	3958	Couteau (1965)			55.88	0.785	1.02	0.5
10075	Σ	2052		7.7–7.8	K2V		.058		
16267	N	1831	Siegrist (1950)			236.07	2.2337	1.86	0.41
10157	Σ	2084	ζ Her	2.9–5.5	G0IV–K0V		.104		N
16394	N	3142	Baize (1975)			34.49	1.355	0.84	0.5
−8°4352	Kui	75		9.8–9.8	M4V–M4V		.161		
16528	S	0814	Voûte (1945)			1.71504	0.218	1.25	–
10374	β	1118	η Oph	3.0–3.5	A2.5V		.051		
17075	S	1540	Knipe (1959)			84.31	1.057	0.57	0.50
+45°2505	Kui	79		10.0–10.4	M3V–M3V		.155		
17106	N	4547	Baize (1952)			12.98	0.71		

162 Computation of Orbits and Stellar Masses

Table 6.11 (cont.)

ADS	α 1950 δ	Name	Computer and Year	Magnitude	Spectrum	P(Years)	p_{tri}	a''	$\Sigma M_\odot$	R
−34°11626	Mlb	0		6.1–7.6	K3V–K5V			.140	1.07	0.41
17156	S	3456	Wielen (1962)			42.177		1.7338		
10585	A	351		9.6–10.0	dM0	60.0		.055	0.36	–
17274	N	2926	Baize (1954)					0.60		
10598	Σ	2173		6.0–6.1	G8IV	46.08		.060	2.31	0.486
17278	S	0101	Duncombe and Ashbrook (1951)					1.02		N
10660	β	962	26 Dra	5.3–7.9	G0V			.073	1.60	0.37
17345	N	6155	Baize (1965)			76.00		1.52		N
10786	AC	7		10.3–10.8	dM4–dM4			.124	0.71	–
17445	N	2545	Couteau (1958)			43.20		1.360		
11046	Σ	2272	70 Oph	4.2–6.0	K0V–K5V	87.892		.195	1.64	0.42
18029	N	0232	Wielen (1962)					4.5482		N
11077	AC	15	99 Her	5.1–8.5	F7V–K5V	55.8		.061	1.41	0.37
18051	N	3033	Heintz (1970)					1.00		
11871	β	648		5.4–7.5	G0V	61.203		.054	3.23	0.43
18552	N	3250	Schrutka-Rechtenstamm (1934)					1.24		
12889	Σ	2576		8.3–8.4	dK5–dK5			.047	1.61	0.5
19436	N	3330	Baize (1954)			243.55		2.15		
+4°4510	Kui	99		8.4–9.1	K5V			.062	1.39	0.489

163 Computation of Orbits and Stellar Masses

20370	N		Baize (1969)			42.35	0.84			
14773	OΣ	535	δ Equ	5.2–5.3	F8V–F8V		.055	3.25	N	0.49
21120	N	0948	Luyten and Ebbighausen (1934)			5.70	0.26			
14787	AGC	13	τ Cyg	3.8–6.4	F0IV		.054	1.74		0.43
21128	N	3749	Heintz (1970)			49.9	0.88		N	
−58°7893	φ	283		9.4–9.5	K7V		.052	2.27		0.5
21430	S	5755	van den Bos (1949)			6.32	0.2335			
15972	Krü			9.8–11.5	dM3–dM4		.253	0.44		0.38
22263	N	5727	Lippincott (1952)			44.6	2.412			
17175	β	733	85 Peg	5.8–8.9	G3V		.084	1.26		0.48
23595	N	2649	Wielen (1962)			26.386	0.8032			

*(N indicates a note in the following.)

Notes

ADS 48   According to Baize (Comm. Coimbre Colloquium 1974) there is probably a third body with a period of 7.3 years.

ADS 61   The mass ratio indicates that there may be a third body, B'. According to Baize the masses are $m_A = 0.9$, $m_B = 0.7$, $m_{B'} = 0.7$. The pair BB' should have a period of 6.9 years. (Comm. Coimbre Colloquium 1974.)

ADS 671   Parallax and mass ratio from K. Aa. Strand, *Astronomical Journal* 74 (1969): 760. Baize suspects two invisible companions of periods 40 and 8.7 years (Comm. Coimbre Colloquium 1974).

L 726-B   Data from C. E. Worley and A. L. Behall, *Astronomical Journal* 78 (1973): 650.

ADS 1865 Parallax and mass ratio from J. F. Wanner, ibid. 74 (1969): 229.

Ross 614 Complete new study by S. L. Lippincott and J. L. Hershey, ibid. 77 (1972): 679.

ADS 6420 Parallax and mass ratio from L. A. Breakiron and G. Gatewood, ibid. 89 (1975): 318.

+42° 1956 Complete new study by L. A. Breakiron and G. Gatewood, ibid. 80 (1975): 714, but table gives the orbit by Baize, which represents the observations better.

ADS 8862	Parallax and mass ratio by W. D. Heintz, ibid. 74 (1969): 768.
ADS 9413	Strand drew attention to a perturbation of period 2.2 years. It was not confirmed by Baize, who found one of 10.2 years. It is probable, therefore, that there is a third body of low mass (Comm. Coimbra Colloquium, 1974).
ADS 9617	New study by L. A. Breakiron et al., *Astronomical Journal* 80 (1975): 174, but table gives the orbit by Danjon, which satisfies the observations better. Baize suspects a third body with period close to 8 years (Comm. Coimbra Colloquium 1974).
ADS 10157	According to Baize, the primary is an astrometric double with a period of 10.5 years and a major semiaxis of 0″.12. The mass of the invisible companion is 0.2 (Comm. Coimbra Colloquium 1974).
ADS 10598	A. H. Batten, J. M. Fletcher, and F. R. West [*Publications of the Astronomical Society of the Pacific* 83 (1971): 149] found a parallax of 0″.060 from observations of radial velocities.
ADS 10660	Parallax and mass ratio from P. van de Kamp and S. L. Lippincott, *Astronomical Journal* 81 (1976): 775.
ADS 11046	Third body suspected but not confirmed.
+4° 4510	Mass ratio from P. J. Morel, ibid. 74 (1969): 245.
ADS 14787	Parallax and mass ratio by W. D. Heintz, ibid. 75 (1970): 848.

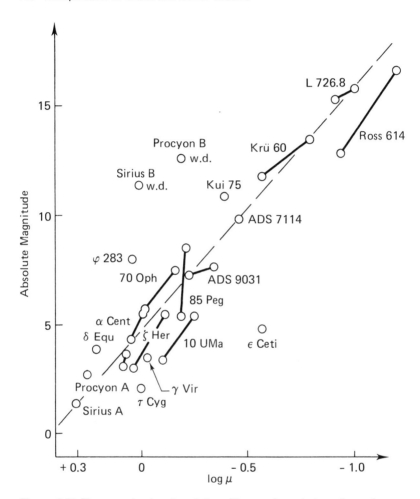

Figure 6.14 The mass-luminosity relation. The number of stars shown has been limited so that the diagram will not be overcrowded. Members of a pair are joined by a line. The dashed line is the mean relation according to Baize and Romani.

close to that defined by all the pairs. The graph suggests that masses and luminosities are not independent. In figure 6.14, the mean line is marked in dashes; it illustrates the mass-luminosity relation found mathematically by Eddington in 1916. A certain number of stars and pairs deviate markedly from this relation. Some of them have too large a mass for their brightness. This is the case for the white dwarfs. Others, to the contrary, have too small a mass, such as ϵ Ceti. Curiously, some companions are more massive than their primaries, as in the systems 9 Puppis and ADS 61.

DYNAMICAL PARALLAXES

The mass-luminosity relation was discovered by Eddington from the study of the internal structure of the stars. Several investigators have put this relation in an empirical form, and have deduced from it a good way to calculate the parallaxes of double stars whose orbit is known. Such parallaxes are called *dynamical parallaxes*; their validity depends on that of the mass-luminosity relation, which the stars obey more or less exactly. The advantage of this kind of calculation is that it makes the estimation of very small parallaxes possible and guides the astronomer in his choice.

There are two principal methods of calculating dynamical parallaxes; that of H. N. Russell and C. E. Moore (1940) and that of P. Baize and L. Romani (1946). We describe the latter, which is easier to use than the former. The two methods, however, give similar results. Baize (1943) wrote the relation in the form

$$\log \mu = -k(M - M_\odot) \tag{6.52}$$

where μ is the mass in solar units, $k = 0.117$, M is the absolute bolometric magnitude of the star, $M_\odot$ is the absolute bolometric magnitude of the sun (equal to 4.77).

We recall that the absolute magnitude is the magnitude a star would have if it were observed from a distance of 10 parsecs. Therefore, the absolute magnitude, the apparent magnitude m, and the parallax p are related by

$$N = m + 5 + \log p. \tag{6.53}$$

This is *Pogson's formula*.

Computation of Orbits and Stellar Masses

The magnitude of a star is a photometric datum related to the wavelengths to which the receiver is sensitive. If $e(\lambda)$ is the energy per unit wavelength, the receiver measures

$$E = \int_{\lambda_1}^{\lambda_2} C(\lambda)e(\lambda)d\lambda.$$

This is the transmission factor of the receiver. If the receiver is the eye, the magnitude is visual; it could be photographic, photovisual, or infrared, according to the plates or photoelectric cells chosen. A magnitude related to the total radiation

$$E_T = \int_0^\infty e(\lambda)d\lambda$$

is called *bolometric*. The difference between the visual magnitude and the bolometric is called the *bolometric correction*; its theoretical value is

$$\Delta m = 2.5 \log(E/E_T) = C.$$

The visual and bolometric magnitudes are related by

$$m_b = m_v + C_v. \tag{6.54}$$

The correction C_v is a function of the spectral type (or temperature) of the stars. Table 6.12 is taken from Harris (1963) for the main sequence, except for type M, for which Baize's values (1943) are given.

Consider a double-star system whose orbit is known, but whose parallax is not. It is easy to show that the combined application of

Table 6.12 Bolometric Corrections.

Spectral Type	C_v	Spectral Type	C_v	Spectral Type	C_v	Spectral Type	C_v
B5	−1.39	A1	−0.32	F5	−0.04	K3	−0.35
B6	−1.21	A2	−0.25	G0	−0.06	K5	−0.71
B7	−1.04	A3	−0.20	G5	−0.10	K7	−1.02
B8	−0.85	A5	−0.15	G8	−0.15	M0	(−1.22)*
B9	−0.66	A7	−0.12	K0	−0.19	M2	(−1.43)
A0	−0.40	F0	−0.08	K2	−0.25	M4	(−1.67)

* Parentheses indicate that a value is relatively uncertain.

equations (6.40), (6.52), and (6.53) makes it possible to calculate a parallax and values of the masses which are called the *dynamical parallaxes and masses*. In fact, we write first

$$\mu_A + \mu_B = \mu_A(1 + \mu_B/\mu_A).$$

By equation (6.52) we have

$$\mu_B/\mu_A = 10^{-k\Delta m}, \qquad (6.55)$$

where Δm is the difference in bolometric magnitude between the components (its value is always positive). From equation (6.54) we have

$$\Delta m_b = m_{vB} - m_{vA} + C_{vB} - C_{vA}.$$

If the two spectra are identical, no correction is needed. From equation (6.55) we obtain

$$\mu_A + \mu_B = \mu_A(1 + 10^{-k\Delta m}) = a''^3/p^3 P^2 \qquad (6.56)$$

or

$$\mu_A + \mu_B = \mu_A D = \alpha^3/p^3,$$

where for simplicity we have written

$$D = 1 + 10^{-k\Delta m}$$

and

$$a''^3/P^2 = \alpha^3.$$

Substituting Pogson's law in the mass-luminosity relation and solving for the parallax, we easily obtain

$$(1 - \tfrac{5}{3}k)\log p = \log\alpha + \tfrac{1}{3}k(m_{vA} + C_{vA} + 5 - M_\odot) - \tfrac{1}{3}\log D. \qquad (6.57)$$

This is the formula found by Baize and Romani, but it is derived here in a more direct way. Using the values given by these authors, we obtain

$$\log p = 0.010 + 1.229 \log\alpha + 0.046(m_{vA} + C_{vA}) - 0.410 \log D. \qquad (6.58)$$

Thus the so-called dynamical parallax is obtained as a function of the major semiaxis, the period, the apparent visual magnitude of component A, and the difference in magnitude between the components. These parameters do not all have the same influence on the parallax. Thus, the last term varies from -0.12 for $\Delta m = 0$ to -0.07 for $\Delta m = 3$. The spectral type plays a role through the bolometric correction C_v (which, however, is multiplied by a small factor in the equation, so that the various scales for this correction do not change the result much). The same is true of the influence of different spectra on the function D. The orbital parameters play the determining role.

For those who do not have a calculator, equation (6.58) can be made easier to use by replacing the function $\log D$ by the first two terms of its expansion in series near the origin:

$$\log D = \log 2 - (k/2)\Delta m. \tag{6.59}$$

The masses can be obtained directly from the equations that made possible the solution for the parallax. It is easily found that

$$\frac{3 - 5k}{k} \log \mu_A = 5 \log D - 3 \log \beta - 15 \log \alpha, \tag{6.60}$$

where $\beta = m_{vA} + C_{vA} - M_\odot + 5$, which gives, with the values adopted by Baize and Romani and taking into account equation (6.59),

$$\log \mu_A = -0.0374 - 0.137(m_{vA} + C_{vA}) - 0.686 \log \alpha - 0.0125 \Delta m, \tag{6.61}$$

$$\log \mu_B = \log \mu_A - 0.1120 \Delta m.$$

Thus we obtain the mass, an intrinsic property of the star, without using the dynamical parallax. Computers prefer to obtain the dynamical parallax first, because the quantity can be measured directly, unlike the mass, and because a check is possible when a trigonometric parallax is determined. The application above, however, gives the individual masses, while the dynamical parallax gives only their sum. When the series expansion of $\log D$ is limited to two terms, agreement to better than 1 percent is obtained up to $\Delta m = 2$.

In the equations (6.58) and (6.61) the spectra of the two components are assumed to be known. More often, we know only the mean spectrum. The two spectra are then supposed to be identical. The error thus introduced is negligible, because two very different spectra correspond to a large Δm, and in that case $\log D$ tends to zero. If the spectral type is completely unknown, we can determine, as Baize does, a preliminary mass whose value tells us the probable spectrum, which can be used as a second approximation. Besides the importance of the orbital parameters, that of the visual magnitude should be noted. Good values of parallaxes and dynamical masses must necessarily come from good photometric measures.

Finally, we wish to know the effect of a small variation in the factor k on the masses. Take the partial derivative of equation (6.60) with respect to k:

$$\frac{\partial}{\partial k}(\log\mu_A) = \frac{5}{(3-5k)^2} 3\log 2 - \frac{k(6-5k)\Delta m}{2} - \frac{9}{5}\beta - 9\log\alpha. \quad (6.62)$$

The variation of the mass is independent of k if the terms on the right-hand side add up to zero. A trial shows that this is not far from the case for binaries of type G. For the others, the sum is always small. This explains the success of computations of dynamical parallaxes. They are not sensitive to small deviations from the mass-luminosity relation.

Examples

Let us choose α Centauri, whose trigonometrical parallax is known. The orbit by Heintz (1968) gives first

$a'' = 17.583, \qquad P = 79.920$ years.

Moreover, we have

	Component	
	A	B
Spectrum	G2V	K0V
m_v	−0.04	1.17
C_v	−0.08	−0.19

whence $\Delta m = 1.10$, $\log\alpha = -0.02334$, $\log D = 0.2439$ and we obtain the following comparative table.

	Dynamical	Trigonometric
Parallax	0″.751	0″.743
μ_A	1.14	1.12
μ_B	0.86	0.95

We find a rather small dynamical mass for the companion, conforming to its deviation from the mass-luminosity relation (fig. 6.14).

We now choose ADS 8987 = β 612, of unknown parallax. The orbit by Danjon (1956) gives

$a'' = 0.208$, $P = 22.35$ years.

On the other hand, $m_{vA} = m_{vB} = 6.3$ and the mean spectrum is A6, whence $C_v = -0.13$. Thus, $p_d = 0″.01686$ and $\mu_A = \mu_B = 1.89$.

BIBLIOGRAPHY

These references, given in addition to the works cited in the text, follow the same order as the chapter.

Orbits: Generalities

There are several tables for converting mean anomalies into true anomalies:

Schlesinger, F., and S. Udick. "Tables for the True Anomaly in Elliptic Orbits." *Publications of Allegheny Observatory* 2 (1912). (This table goes up to $e = 0.77$.)

There is a supplementary table, made in response to a request at the Meeting of the International Astronomical Union in Berkeley:

Muller, P. "Table des anomalies vraies jusqu'à l'anomalie moyenne de 20 degrés pour les excentricités égales ou supérieures à 0.75." *Astrometrie* no. 5 (1961): Notes et Informations, fasc. VI.

An important table was published in the USSR in 1960:

Jongolovitch, I. D., and V. M. Amelin. *Summary of Tables and Nomograms for Computing Trajectories of Artificial Earth Satellites.* Institute of Theoretical Astronomy of the USSR Academy of Sciences. (In Russian.) There is also a table for Keplerian motion in Danjon's *Astronomie générale*. The reduced coordinates X and Y have been given as functions of the eccentricity and the mean anomaly in the *Appendix to Union Obs. Circular* no. 71 (Pretoria, 1927).

The use of pocket electronic calculators is making trigonometric tables less and less necessary. All the same, we give a reference to a very convenient work:

Danjon, A. *Fonctions trigonométriques. Valeurs naturelles à six décimales.* Hachette, 1948.

Computation of Orbits (Geometric Methods)

Baize, P. "Comment calculer une orbite d'étoile double." *L'astronomie* 68 (1954): 22.

van de Kamp, P. "Two Graphical Procedures for Evaluating the Excentricity of an Astrometric Double Star Orbit." *Astronomical Journal* 52 (1947): 185.

Computation of Orbits (Method of Thiele, Innes, and van den Bos)

Arend, S. "Etablissement par voie raccourcie des formules de Thiele-Innes relatives aux orbites d'étoiles doubles visuelles, en recourant aux principes de l'affinité." *Ann. Obs. Toulouse* 16 (1941): 109.

———. Le rôle joué par les constantes de Thiele-Innes dans le calcul des orbites des étoiles doubles visuelles. Commun. Obs. Royal de Belgique no. 195 (1961).

———. Détermination du moyen mouvement annuel du compagnon d'une étoile double visuelle dont l'orbite est calculée part la méthode de Thiele-Innes. Commun. Obs. Royal de Belgique no. 243 (1970).

Binnendijk, L. *Properties of Double Stars: A survey of parallaxes and orbits.* University of Pennsylvania Press, 1960.

Dommanget, J. "Propriété du système des équations fodamentales de la méthode de Thiele-Innes pour le calcul d'orbites d'étoiles doubles visuelles." *Journal des Observateurs,* vol. 42, p. 129, 1959.

van den Bos, W. H. "Orbit Determinations of Visual Binaries." In W. A. Hiltner (ed.), *Astronomical Techniques.* University of Chicago Press, 1962.

Computation of Orbits (Method of Opposite Points)

Danjon's publications have been cited in the text.

Muller, P. "Table pour le calcul des éléments des orbites des étoiles doubles visuelles." *Ann. Obs. Strasbourg* 5 (1956).

Morel, P. J. "Remarques sur l'utilisation de la méthode des moindres carrés dans la méthode des points opposés de Danjon pour le calcul des orbites d'étoiles double visuelles." *Bull. Astron. Obs. Royal de Belgique* 7 (1971): 191.

Orbit Catalogues. Ephemerides

Finsen, W. S., and C. E. Worley. Third Catalogue of Orbits of Visual Binary Stars. *Republic Obs. Johannesburg Circ.* 7, no. 129 (1970).

Dommanget, J., and O. Nys. Catalogue d'éphémérides des vitesses radiales relatives des composantes des étoiles doubles visuelles dont l'orbite est connue. *Commun. Obs. Royal de Belgique* no. 15 (1967).

Muller, P., and C. Meyer. Troisième catalogue d'éphémérides d'étoiles doubles. *Publ. obs. Paris* (1969).

Masses and Mass Ratios

van de Kamp, P. "Long focus photographic astrometry." *Popular Astronomy* 59, nos. 2–5 (1951).

van de Kamp, P. *Principles of Astrometry.* San Francisco: Freeman, 1967.

Morel, P. J. Contribution a la determination photographique du rapport de masse d'une binaire visuelle. Thèse de Spécialité, Nice Observatory, 1969.

Muller, P. Etoiles doubles visuelles. Masses des étoiles. *Annuaire du Bureau des Longitudes,* 1970.

Dynamical Parallaxes

Barbier, D. "Les Parallaxes dynamiques des étoiles doubles." *Actualités Scientifiques et Industrielles,* (1936): 348.

Russell, H. N., and C. E. Moore. *The Masses of the Stars.* University of Chicago Press, 1940.

Baize, P. "Les masses des étoiles et la relation empirique masse-luminosité." *L'Astronomie* 57 (1943): 101.

———. "Les masses des étoiles doubles visuelles et la relation empirique mass-luminosité." *Bulletin Astronomique* 13 (1947): 123.

Baize, P., and L. Romani. "Formules nouvelles pour le calcul des parallaxes dynamique des couples orbitaux." *Ann. d'Astrophys.* 9 (1946): 13.

Dommanget, J. "Les Parallaxes dynamiques." *Ciel et Terre* 72 (1976): 65.

Couteau, P. "Sur la validité de la relation mass-luminosité dans le calcul des masses des étoiles." *Astrophysics and Space Science* 11 (1971): 55.

Harris, D. L. III, K. A. Strand, and C. E. Worley. "Empirical Data on Stellar Masses, Luminosities and Radii." In *Basic Astronomical Data.* University of Chicago Press, 1963. (The authors of this article give two mass-luminosity relations: $M_b = 4.6 - 10.0 \log \mu$, valid for stars brighter than $M_b = 7.5$, and $M_b = 5.2 - 6.9 \log \mu$, for all other stars. The relation given by Baize is intermediate between these, and has the advantage that the same relation can be applied for all masses.)

Baize, P. "Les Étoiles doubles et la relation masse-luminosité." *L'Astronomie* 89 (1975): 306.

7
A VOYAGE TO THE COUNTRY OF DOUBLE STARS

SOME TYPICAL SYSTEMS

Looking at a system of stars through a telescope transports the mind on a distant voyage. If the system is at a known distance and is composed of objects that the astronomer can compare with our sun, it is easy to travel there in thought and to imagine the sights seen by the inhabitants of that place. Often, the system will not give up its secret; neither the distance nor the orbit is known. That is the most usual situation. Almost all the double stars that have been classified are of this kind; they are formed of very luminous objects, circling slowly in tremendous orbits. Very many appear fixed after nearly two centuries of observation, such as the companions of Arcturus or Rigel.

By contrast, the well-known systems form a family from which we learn a great deal, and which, thanks to our knowledge of dynamical parallaxes and masses, gives us some idea of the stellar population in the neighborhood of the sun.

There are double stars of all possible periods, from a few hours for stars in contact (unobservable visually, but not spectroscopically) up to thousands of years. At very large separations, perturbations can dissociate the system so that each component becomes a single star. Among the 700 double stars with known orbits are to be found all the possible orbital eccentricities and inclinations. All types of stars are represented. It appears that binaries do not constitute a privileged class; nothing in the structure of a star differentiates a single star from a member of a double. The number of binaries is so great that the majority of objects belong to a system; single stars like our sun are a minority.

The distribution of double stars in the sky is the same as that of single stars; it is galactic. The same is not true for the orbits, which are distributed homogeneously. This means that pairs with known orbits occupy a volume that is small compared with the galaxy. Dynamical parallaxes show that these systems are situated within a sphere of about 100 parsecs radius, centered on the sun.

The study of double stars enables us to compare stars with the sun in mass, luminosity, and size. Known systems are comparable in size with the solar system as far out as Pluto. We recall that our great luminary is in the middle of both the mass and luminosity scales. Volumes of stars differ so much that it is difficult to be precise about them. There are stars as massive as the sun, but smaller than a big city; others are so large that they could not fit inside the solar system. Red dwarfs are plentiful; they are the most numerous population. Among the hundred nearest stars, the sun is fourth in radiative power. Surface temperatures range from 2,500°K for type M to tens of thousands of degrees for stars of types O and B.

It is not difficult to imagine the sight that would be presented to us if our earth were a planet of a double star. What Lilliputians or Brobdingnagians can we imagine as inhabitants of an earth circling around Sirius, 2.5 parsecs or 8 light-years away, at the same distance as we are from the sun! Sirius, a bright white sphere over a degree in diameter, would shine 100 times as bright as our sun. The companion of Sirius would appear at night like a single star, an arc-second in diameter, shining with a white light like that of our full moon in intensity. This white dwarf has a diameter of 15,000 km despite the fact that its mass is comparable with that of the sun. Viewed from here, this small star can be hidden by the thickness of a hair 2,000 km away. The surface of this object would be indiscernible; it is probably of uniform smoothness, since the force of gravity, 2,000 times ours, forbids any structure or relief whatsoever.

In our nearest neighbor, α Centauri, 4 light-years away, we find another star like the sun; nothing would appear to us to be changed by day, but our nights would be brightened by the beautiful copper-colored companion, revolving in an orbit as large as that of Neptune. This companion would not appear to our naked eyes to have an appreciable diameter, but it would be unbearably bright, shining

176 Voyage to the Country of Double Stars

with a light 1,000 times that of the full moon. Sometimes, near opposition, there would be no nights; the two suns would take turns above the horizon (fig. 7.1).

Several systems have more than two stars and form a dynamical hierarchy like the sun-earth-moon system. Thus, the inhabitants of the system Kui 23, 71 light-years away, are lit by a central object resembling Capella, 50 times bigger than the sun. Their nights are illuminated by a sun that is itself double, revolving in an orbit intermediate between those of Jupiter and Saturn. This satellite system is formed of two stars, indistinguishable to us, roughly the size of our sun, turning one about the other in 9 days, at a mutual distance of 7 million km. These beautiful red twin objects eclipse each other from time to time and make a natural clock in the skies of the inhabitants of Kui 23; they would inspire painters and poets, and give the astronomers of the place inexhaustible subjects of research.

The inhabitants of the triple system 13 Ceti, 56 light-years away,

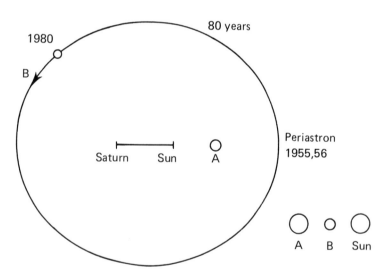

Figure 7.1 True orbit of α Centauri. The system contains a third very distant red dwarf star. The star A is comparable to our sun, and the star B revolves around it in an orbit whose size is comparable to that of Neptune. The circles to the lower right represent the relative sizes of the components and of our sun.

177 Voyage to the Country of Double Stars

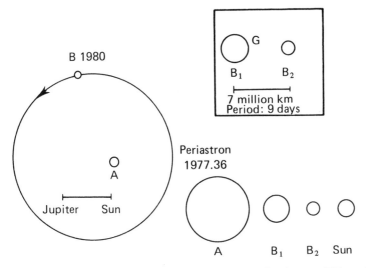

Figure 7.2 True orbit of the triple system Kui 23 = 1 Geminorum. This system comprises a single primary star and a double companion $B_1 B_2$ revolving in 13 years in a slightly eccentric orbit about the size of that of Saturn. (In the square, the system of B, which is not a visual binary, is represented on a 100× scale.) The stars B_1 and B_2 revolve around their center of gravity, G, in 9 days, in a circular orbit 7 million km in diameter. At the bottom of the picture, the sizes of the stars are compared with that of the sun.

would have the benefit of a double central body formed of two dissimilar stars, yellow and red, revolving, 5 million km from each other, in 2 days. The eternal waltz of these suns would also inspire artists and astronomers. The nights are lit by the red companion, which every 7 years would approach the point of rivaling the principal star in brightness. It would be an enchantment of light, since sometimes there would be no night; the dazzling red sun would rise as the double one set, touching with two colors everything that surround the people there. At other times, three suns would light the countryside and cast three shadows of each object.

In some other systems, such as ξ Ursae Majoris, each component of the double star is itself double. Castor is sextuple; the visual pair is formed of two double components, and around this system, at a great distance, a pair of red dwarfs is also turning.

178 Voyage to the Country of Double Stars

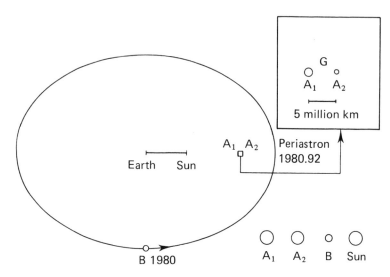

Figure 7.3 True orbit of the triple system ADS 490, 13 Ceti. This system comprises a primary A_1A_2, itself double but not a visual binary, and a companion B revolving about it in 7 years in a fairly eccentric orbit a little smaller than that of Jupiter. In the square, the system A_1A_2 is shown, with its center of gravity, on a 20× scale. The period of system A is about 2 days. At the bottom, the sizes of the stars are compared with that of the sun.

These few examples show that visual double stars are composed of objects at large distances, some hundreds or thousands of times their diameters, from each other. This is true even of those whose components seem to touch each other: Diffraction images are much larger than the objects themselves, which have apparent diameters of the order of a thousandth of an arc-second.

All these systems are hierarchical, unlike star clusters in which the center of attraction is the cluster itself. Stars in clusters do not move in mutual orbits; they belong to systems like the Pleiades, in the process of dissociation, or, like the globular clusters, in equilibrium. Double stars afford the sight of Keplerian motion about several centers of gravity. Other examples are the satellites of the solar system. For example, artificial satellites of the moon obey the force of gravity of the moon, which itself gravitates around the barycenter of the earth-moon system. (Here we are skimming over difficult problems of celestial mechanics, good examples of whose applications are provided by some triple stars at the limit of dynamical stability, such as ζ Herculis.)

BENEFITS OF THE VOYAGE AS A FUNCTION OF THE INSTRUMENT

The observer really makes a voyage into the country of double stars, since he sees the heavenly bodies as if he had drawn closer to them. Some choices are possible, however: both light-gathering power and resolution are gained with a bigger instrument. To see the bustling universe, to count the stars, to walk down the Milky Way, to discover objects that no one else has seen before—all these depend on the size of the instrument. Let us specify the possibilities of the voyage by emphasizing a number of its aspects. Like travel agents, we will make you some offers according to your means.

The first question to study is: How long must we wait to know the orbit of a double star that has just been discovered? The answer to this question defines the *rate of increase* of information. To be able to calculate an orbit, an interval of about half an orbital period must elapse from the discovery of the object. How long this is depends on the separation of the stars and the aperture of the instrument.

When a pair is discovered, the separation ρ is measured. Statistically, the major semiaxis can be supposed to be little different. The author showed some years ago (1960) that we can write

$a'' \approx 1.25\rho.$

The astronomer who looks for the most rapidly moving pairs will select those that are at the limit of resolution. Thus, ρ is given by the resolving power $12/D$, which enables us to deduce the major semiaxis:

$$a'' \approx 15/D, \quad \text{or} \quad a_{AU} \approx 15r/D, \tag{7.1}$$

where r is the distance of the system in parsecs and D is the aperture in centimeters. On the other hand, we can assume that the mass of the system is close to twice that of the sun. By Kepler's third law, this gives an order of magnitude for the period:

$$P^2 \approx 1.2 \times 10^3 r^3/D^3. \tag{7.2}$$

180 Voyage to the Country of Double Stars

For a given parallax, the shortest periods P_1 and P_2 accessible with apertures D_1 and D_2 obey the relation

$$P_1/P_2 = (D_2/D_1)^{3/2}. \tag{7.3}$$

If the aperture is doubled, the shortest accessible period is smaller by a factor 2.8, at the same distance. Thus, information increases as $D^{3/2}$.

Table 7.1 gives the rate of increase of information for several apertures looking to a distance of 100 parsecs. This table shows the advantage of large apertures for discovering binaries and computing their orbits without undue delay. For an astronomer to have a chance of knowing the orbits of the binaries he discovers, he should have the use of a refractor of at least 70 cm aperture. Even so, he would have to wait 25 years, on average, for objects 100 parsecs away.

PENETRATING POWER OF AN INSTRUMENT

Let us study another question: To what distance in space, and with what instrument, can pairs with periods less than a given value be seen? The answer to this question gives the *orbital range* of an instrument. To answer it simply, it is useful to make the following restrictions:
- Consider only periods less than 200 years, corresponding to an information time of one century, which is already a long delay—quite a bit longer than an astronomer's active life.
- Limit consideration to pairs that can be resolved by the instrument.
- Suppose that the masses are always similar to that of the sun.
- Limit consideration to pairs containing equal components, each

Table 7.1 The Growth of Information at a Distance of 100 Parsecs.

D (cm)	P (years)	Information Time (years)
150	18	9
100	35	17
50	97	48
25	280	140
10	1,100	550

of which is not fainter than magnitude 10. Experience shows that, whatever the aperture, magnitude 10 is a barrier. In a large instrument the images lose their sharpness and break up. Light is lost in the diffraction rings, and the eye does not receive very much more illumination.

The first and third conditions limit the major semiaxes to $a_1 = 43$ AU. The second condition imposes a lower limit below which the components can no longer be separated, namely,

$$a_r \approx 15r/D \tag{7.4}$$

in astronomical units. The fourth condition, on the other hand, gives the relation

$$\log r \approx 3 - 0.2M, \tag{7.5}$$

which defines the radius of the sphere, centered on the sun, in which stars of absolute magnitude M are at least as bright as magnitude 10. Combining this last relation with $a_1 = 43$ AU, we deduce the *maximum apparent* major semiaxis a_1'', above which the period is more than two centuries:

$$\log a_1'' \approx 0.2M - 1.37. \tag{7.6}$$

Finally, applying Pogson's law to equation (7.4), we find the *minimum* major semiaxis in astronomical units, a_r:

$$\log a_r \approx 3 - 0.2M + \log(15/D). \tag{7.7}$$

When $a_r = a_1 = 43$, that is, when

$$\log r \approx \log D + 0.45, \tag{7.8}$$

we have reached the instrumental limit for the observation of orbits, so equation (7.8) defines the *orbital range*. This limiting distance is associated with the absolute magnitude

$$M \approx 12.75 - 5\log D. \tag{7.9}$$

These relations give orders of magnitude. One could refine them, at the cost of some complication in writing them, by using the mass-luminosity relation. The answer sought, however, would not be made any clearer.

We can summarize these few considerations in a table (see table

Table 7.2 Penetrating Power of an Instrument.

	Spectral Type					
	A5	F5	G0	G5	K0	K5
M^a	2.2	3.5	4.4	5.1	6.0	7.8
r (parsecs)[b]	360	200	130	96	63	28
V^c	47	8	2	0.9	0.25	0.02
$a_1''{}^d$	0''.12	0''.21	0''.32	0''.45	0''.67	1''.55
D (cm)[e]	125	70	45	33		

a. Absolute magnitude.
b. Distance at which star would be of apparent magnitude 10.
c. Corresponding volume of sphere of radius r parsecs, in units of the volume of a sphere of radius 100 parsecs.
d. Apparent separation for which period would be 200 years.
e. Corresponding limiting aperture.

7.2, which is limited to stars on the main sequence) and illustrate them by a diagram. Later than type G5, it is no longer diffraction but brightness that limits observation of red dwarfs. With an aperture of 125 cm, periods of two centuries can be observed out to 360 parsecs, or 1,200 light-years, for stars of type A5 or cooler. The volume explored is 24 times larger than that corresponding to an aperture of 45 cm. Under the same conditions, this latter aperture can reach solar-type stars as far away as 130 parsecs. Of course, double stars that are much farther away and much brighter can be observed, but their periods exceed two centuries, and the computation of their orbits must wait for some time.

The country of double stars with computable orbits has its frontier about 1,200 light-years away, with our current means of observation. This is already a considerable volume; it takes up the whole thickness of our part of the Galaxy, which is far from the center. Table 7.2 shows why stars of types F and G are favored by observational selection; the limiting pairs associated with them correspond to apertures from 40 to 90 cm. These are precisely the dimensions of the large refractors used in the surveys. These pairs are situated between 100 and 300 parsecs from us.

Figure 7.4 delimits the region in which orbits are observable. The horizontal line corresponds to the maximum separation beyond which periods are greater than two centuries. The lines passing through the origin determine the minimum major semiaxes for in-

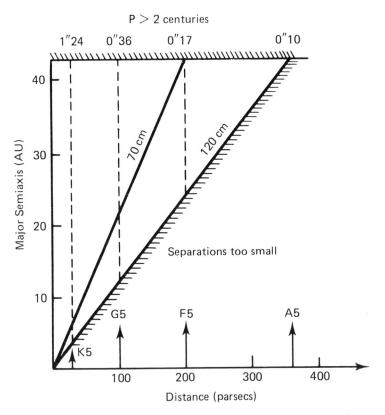

Figure 7.4 With an aperture of 120 cm, orbits with periods less than two centuries can be observed as far away as 350 parsecs, or 1,150 light-years. At that distance, however, only giants can be observed. Solar-type stars are observable only as far as 150 parsecs, or 500 light-years. With a 70-cm refractor, a distance of only 200 parsecs, or 650 light-years, can be reached.

struments of 120 and 70 cm aperture. The vertical dashed lines indicate the limits of distance beyond which stars of the types indicated by the arrows are too faint for observation.

Examples

The orbit of an F5 star is observable up to 200 parsecs; beyond that it is too faint ($m > 10$). At the limit of 200 parsecs, separations must lie betwen $0''.1$ and $0''.17$.

The orbit of a B-type star is observable as far away as 360 parsecs.

Imagine a pair of G5 stars, 160 parsecs away, whose orbit has a major semiaxis of 30 AU. Its separation would permit its orbit to be observed, but not its magnitude.

Imagine, now, a pair of A5 stars, 300 parsecs away, with an orbit of major semiaxis 20 AU. It is bright enough for us to see it, but its separation is too small for us to compute its orbit.

Thus, the large relative proportion of red dwarfs with known orbits, in relation to other types, is explained. All the red dwarfs recorded are necessarily close to us. A given angular separation corresponds to a smaller apparent major axis for them than it does for the other types, hence the short periods that become known fairly quickly. Whereas the hot B-type stars are, on average, much farther away, their angular separation is visible only if it corresponds to a long-period orbit.

NUMBER OF ORBITS ACCESSIBLE TO AN INSTRUMENT

We will now estimate more precisely the distribution law of orbits according to the spectra of the stars. Let

$$dn = q(a)da \tag{7.10}$$

be the number of binaries per unit volume with semiaxes between a and $a + da$, at a distance r from the sun. We wish to find the number of orbits that we can hope to observe with an aperture D, as a function of spectral type.

The number of observable binaries at the distance r is

$$n(r) = \int_{a_r}^{a_1} q(a)da = q(a_1 - a_r),$$

185 Voyage to the Country of Double Stars

if q is supposed constant, which is reasonable since the interval of integration is small. For each spectral type, the constant q can be obtained from the stellar density in the neighborhood of the sun and from the observed proportions of binaries. The number of observable binaries, N_{sp}, for each spectral type, can therefore be written

$$N_{sp} = \int_0^{r_0} 4\pi q r^2 (a_1 - a_r) dr. \tag{7.11}$$

The distance r_0 is limited for the brightest types by

$\log r = 0.45 + \log D,$

since the magnitude limit gives large distances for these stars. For the fainter types, r_0 is limited by the magnitude limit of 10 through the relation

$M = m + 5 - 5 \log r.$

With these conditions, the number N_{sp} can be computed for each type A, F, G, K, and M of class V. Applying (7.4), we find for bright stars

$$N_{sp} \approx \frac{\pi q a_1^4 D^3}{3 \times 15^3}, \tag{7.12}$$

which shows that the amount of information received increases as the cube of the aperture. For fainter stars we have

$$N_{sp} \approx 4\pi q r^3 (a_1/3 - 15rD/4). \tag{7.13}$$

Table 7.3 shows, for different apertures, the number of binaries with observable orbits and the number actually known. Note the differences between those that are known and those that are observable, and that type M is better represented than would be expected. This

Table 7.3 Numbers of Binaries Accessible to an Instrument.

	Spectral Type					
D	A	F	G	K	M	All
100 cm	1,400	1,660	700	230	9	4,000
40 cm	175	230	343	189	9	950
Known	66	139	106	50	21	382

can be explained by the fact that M-type binaries are widely separated, often by several arc-seconds, so their orbits are easier to obtain and the magnitude limit for them is reduced to about 12.

Thus, there are effects of observational selection that can be recognized and allowed for in calculating the ratio R of the number of binaries observable to that of all pairs, for each type. Let us return to equation (7.10) and write the number of pairs whose orbits are observable; the summation is made between a_r and a_1, then extended to the sphere corresponding to the limiting magnitude, 10, with $a_1 D/15$ [equation (7.1)] as the upper limit:

$$\log r = 3 - M/5 \leq \log(a_1 D/15).$$

We have

$$n(M) = \int_0^r \int_{a_r}^{a_1} 4\pi \, q(a) r^2 dr da$$

$$= \tfrac{4}{3}\pi r^3 Q(a_1) - \int_0^r Q(a_r) 4\pi r^2 dr.$$

On the other hand, if we take account of all pairs with major semiaxes in the interval D to a_1, we have

$$N = \int_0^r \int_0^{a_1} 4\pi \, q(a) r^2 dr da = \tfrac{4}{3}\pi r^3 Q(a_1).$$

The ratio sought, R, is $n(M)/N$, and can be written

$$R(M) = 1 - 3 \frac{\int_0^r Q(a_r) r^2 dr}{r^3 Q(a_1)} = 1 - F(r) \tag{7.14}$$

with the condition $r \leq a_1 D/15$. With the opposite condition, the ratio R becomes

$$R(M) = [1 - F(a_1 D/15)] \frac{a_1^3 D^3}{15^3 r^3}. \tag{7.15}$$

By supposing the function q to be constant (a hypothesis with little justification, but reasonable in view of our ignorance of the distribution law of major axes), we find

$$\log(1 - R) = 2.05 - \log(a_1 D) + (m - M)/5 \quad \text{for } r \leq a_1 D/15,$$
$$\log R = -7.13 + 3 \log(a_1 D) + 0.6(M - m) \quad \text{for } r > a_1 D/15.$$

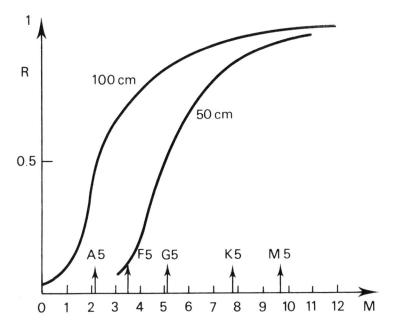

Figure 7.5 Orbital yield of an objective. The figure shows the fraction of observable pairs of periods less than two centuries, as a function of spectral type. With an aperture of 1 m half of the systems containing A5 stars can be observed, but scarcely a tenth of them are observable if an aperture of only 50 cm is available.

For fainter pairs, not visible from a great distance, R tends to unity. It decreases quickly as one goes to hotter stars. We can know the orbits of only a quarter of the pairs of type A, even with the largest apertures. Figure 7.5 shows the ratio R for apertures of 1 m and 50 cm. In summary, three-fourths of the double stars of type A, of major semiaxis less than 40 AU, and brighter than or equal to magnitude 10, are too close to each other to be observable. This fraction becomes one-half for type F and one-fourth at type G7. When a tenth-magnitude double star of type B is discovered, there is practically no chance that it will have an observable orbit. When an M-dwarf pair is discovered, one is sure to obtain an orbit fairly quickly.

PERFORMANCE OF THE EYE COMPARED WITH ITS COMPETITORS

A photographic emulsion is more sensitive than the eye, but the latter can study a diffraction image better. We can compare the performance of the eye with those of its competitors, and rank it among the numerous detectors now available (fig. 7.6). To simplify the comparison, we choose the case of a pair of equal components of solar type, with a period of two centuries, which is at the limit of the resolving power, s. The apparent magnitude of the components will be

$m = 7.9 - 5 \log s$.

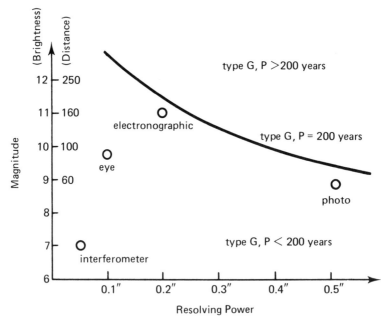

Figure 7.6 The limits of different kinds of receivers in resolution and sensitivity. The curve marks the position occupied by orbits of solar-type stars with periods of two centuries. The eye has good resolving power; it is more sensitive than interferometers.

189 Voyage to the Country of Double Stars

Table 7.4 Comparison of Detectors.

	Sensitivity	Resolution	Yield	Handling of Information
Eye	Low	Good	High	Not possible
Photographic plate	High	Low	High	Possible
Electronographic camera	High	High	Poor	Possible
Interferometer	Low	High	Poor	Possible

This relation delimits the region in the (m,s) diagram in which binaries have periods less than 200 years. On this diagram, the points representing the current performances of the eye, the interferometer, the photographic plate, and the electronic plate are marked. From it we deduce the summary given in table 7.4.

THE VOYAGE OF THE ASTRONOMERS FROM NICE ON BOARD THEIR ECONOMICAL SPACESHIP

The best-equipped observatory in France for the study of double stars is on the wooded hill of Mont-Gros, to the east of Nice. The two big refractors of 74 cm and 50 cm enable the astronomers to travel from one star to another. Each auspicious night, a purely visual expedition takes the astronomer hundreds of light-years to where he can walk among objects that no one else has seen before. There are no mileage charges, accidents are rare, and he always returns, even if he loses himself among the stars. Driving this wonderful space engine, which is a refractor, can be tiring. To stand eight to ten hours with one's arms in the air, one's neck twisted and legs bent, in order to examine 600 or 700 stars with magnifications of 600 to 1,000, requires an athlete's training—excellent for the health. As is the case for all astronauts, the voyage must be prepared carefully; nothing can be left to chance. Discoveries are only the reward for the punctilious execution of a minutely elaborated plan.

As was said in the first chapter, Paul Muller and I are surveying the northern hemisphere: Muller to the north of declination 55°, and I from 17° to 55°. The survey began about 10 years ago. Muller has done almost all his part of the work, harvesting 600 binaries; I have discovered 1,500 pairs in 6,500 square degrees, among the 11,000 stars that make up the program. Muller surveyed stars in the AGK 3 catalogue, while I examined those in Argelander's catalogue. The stars in this latter compilation have been filed on magnetic tape, with whose aid lists convenient for observation have been drawn up by zones of 2°. Pairs already known are noted with the help of the magnetic-tape version of the *Index Catalogue.* They serve as landmarks in the sky; like signposts, they indicate the best routes.

Table 7.5 shows what we have contributed in the area of the sky we have observed. For small separations, the number of new pairs is greater than the number previously known. It is precisely these close pairs that we are looking for; the short-period orbits are to be found among them. Besides, many double stars play hide-and-seek with observers, looking single at the time of the survey only to appear double some time later. This is particularly the case with θ Coronae Borealis, τ Arietis, 13 Pegasi, and 39 Comae, all easily visible now but not so at the beginning of the century. Amateurs, modestly equipped, could have discovered these stars before the professionals.

A voyage is only of value if it proves an enriching experience. Discovering double stars is not an end in itself; we must use the discoveries to increase our knowledge of the laws that govern the universe.

A total of 2,000 new pairs is significant enough to permit conclusions to be drawn. First, we will sort the pairs into giants and dwarfs, or, to put it another way, separate them by luminosity class. Then we will use these results to estimate the period of a binary as soon as it is discovered. Finally, we will give a list of interesting pairs that should be kept under constant surveillance.

Table 7.5 The Survey from Nice Observatory.

	Area Observed by Couteau		Area Observed by Muller	
ρ	Pairs already known	New pairs	Pairs already known	New pairs
$\leq 0\rlap{.}''2$	100	152	35	87
$< 0\rlap{.}''5$	450	472	200	244
$< 1''$	800	854	400	350

191 Voyage to the Country of Double Stars

SORTING NEW DOUBLE STARS INTO GIANTS AND DWARFS

Among the 2,100 pairs discovered at Nice in about 15 years, and some others besides, several are of long periods and unknown spectral types. We should eliminate from this sorting those that provide no useful information, that is to say the giant stars visible very far away, which form pairs practically fixed on the sky background. It is obvious that, on average, close pairs have the shorter periods. On the other hand, knowledge of the spectral types and proper motions is fundamental to all critical studies. Therefore, we keep those pairs that are closer than 0".5 for which we know the spectrum and the proper motion.

If we plot the frequency distribution of the proper motions for each type of star, we find that it resembles the Gaussian distribution, with a smaller dispersion for the solar-type stars. Most stars have proper motions in the neighborhood of a value characteristic of their spectral type, as shown in table 7.6. Statistically, proper motion is proportional to parallax, so the ratio of these two quantities is nearly constant. Note that, among the binaries discovered, the closest are of type G (or solar type); then come those of types F, K, and A. If the distance of those of type G is taken as unity, then the distances of the F-types, the K-types, and the A-types are 1.8, 2.2, and 2.8 respectively. Now, the mean apparent magnitude is approximately the same for all pairs discovered. It is possible, therefore, to deduce the differences in their absolute magnitudes, and to draw conclusions. The results are shown schematically in table 7.7. Note that the F-types have the same representation as the A-types, and that the main sequence (class V) is overrepresented among the G-type stars but underrepresented among the K-type stars because K-type giants are of high luminosity and so are visible from a great distance.

Table 7.6 Mean Proper Motions.

	Type			
	A	F	G	K
$\log\mu$ (annual proper motion)	−1.90	−1.70	−1.45	−1.80

With the help of these results, we will show that main-sequence objects can be separated from others. Let us make the working hypothesis that all new doubles are on the main sequence and plot a graph with the logarithm of the proper motion ($\log\mu$) on the abscissa and the logarithm of the parallax ($\log p$) on the ordinate (fig. 7.7). On this graph, the stars are not distributed at random. There is, first of all, a division according to parallax; since the apparent magnitudes are close (between 7.5 and 10), each type occupies a fairly narrow horizontal band between two ordinates. Note that the A-type stars occupy a region high up and to the right, corresponding to small values of parallax and proper motion. Objects of type F and G are in a large region stretched out in proper motion. The K-type stars are in two distinct regions: one corresponding to the main sequence, characterized by large proper motions, and the other obviously occupied by giants. The straight line

$\log p = \log\mu - 0.7$

passes through the zones in which the main-sequence stars are found. On the other hand, the line

$\log\mu = -1.8$

also passes close to many stars that, apart from the very hot objects, do not belong on the dwarf branch. Of course, small proper motion is not an infallible criterion for distinguishing main-sequence stars, but it is a statistical one.

Some objects in this diagram are correctly placed, namely, those of the dwarf branch; they are distributed along the inclined line. The greater frequency of objects on the right of the diagram indicates the presence of giant stars of small proper motions; their absolute visual magnitudes are between 0.5 and 3. This graph enables us, therefore, to determine which stars are on the main sequence. Results for 351 pairs closer than 0".5, whose proper motions and types are known, are presented in table 7.8. Only a quarter of the new binaries of type K belong to the solar luminosity class; this ratio is close to 1/2 for types F and G. It is important to note, however, that those main-sequence M-type and K-type stars that are observed as visual doubles presumably have rapid proper motions correspond-

193 Voyage to the Country of Double Stars

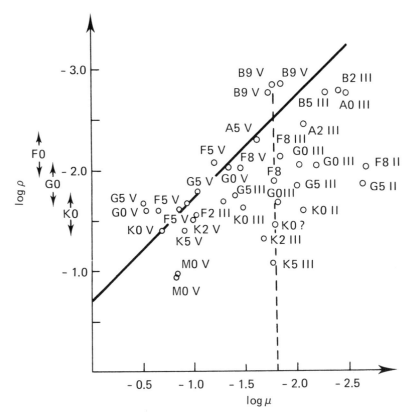

Figure 7.7 Proper motion versus parallax. The diagram represents a number of stars of known proper motion, on the hypothesis that they belong to the main sequence. The stars that really belong to it cluster around the solid line; the others scatter fairly loosely about the dashed line. In this way, the giants (classes III and IV) can be sorted out from the dwarfs (class V).

Table 7.7 Mean Luminosities.

	Type			
	K	G	F	A
Distance	2.2	1	1.8	2.8
Absolute magnitude[a]	$M + 0.50$	$M + 2.25$	$M + 1$	M
Mean absolute magnitude, classes III–V[b]	2.50	2.90	2.13	1.2

a. Relative to that of the A-type group.
b. Computed value, with classes III–V assumed to be represented equally in each sample.

Table 7.8 Properties of 351 Pairs.

	Type					
	K	G	F	A	B	M
Annual proper motion[a]	0.″050	0.″025	0.″016	0.″01	0.″006	0.″12
No. of stars in class V[b]	11	36	69	67	13	1
No. of stars in class IV[b]	11	31	32	13		
No. of stars in other classes	21	12	21	7	(5)	1
Proportion of main-sequence stars	0.26	0.46	0.57			

a. Limiting annual proper motion corresponding to dwarf series.
b. Probable.

ing to their probably large parallaxes. The orbits of these new binaries will be known before those of the others.

PROBABLE PERIOD OF A DOUBLE STAR

It is easy to specify the probable period of a pair by writing an equation that gives it as a function of the quantities observed at the time of discovery. In equation (6.57), let us replace the parallax by its expression as a function of the absolute and apparent magnitudes, and solve for the period. We find that

$$\log P = \tfrac{3}{2}\log a'' + M_A(k/2 - 0.3) - 0.5\log D$$
$$+ 0.3(m_{vA} + C_{vA}) - (k/2)M_\odot + \tfrac{3}{2},$$

or, again, replacing the symbols by their values (chapter 6),

$$\log P \approx 1.5\log a'' - 0.244M_A + 0.3(m_{vA} + C_{vA}) + 0.028\Delta m + 1.08. \tag{7.16}$$

This equation gives the period as a function of the observed quantities and the apparent major semiaxis, provided the star obeys the mass-luminosity relation. As we are interested only in orders of magnitude, we relate the apparent major semiaxis to the separation at the time of discovery by

$$a'' = g\rho$$

and write

$$\log P \approx 1.5\log\rho - 0.244M_A + 0.028\Delta m + 0.3(m_{vA} + C_{vA}) + h, \tag{7.17}$$

where

$$h = 1.08 + 1.5\log g.$$

The value of h varies during the period. We can calculate its mean value for each system; it is not very far from 1.5. Averaged over all binaries, this mean value depends on the frequency distribution of the orbital elements, especially the eccentricity and the inclination. Since these distributions are not well known, we determine h by applying equation (7.17) to double stars with known orbits. However, these double stars are the result of selection, which modifies the mean value of h, so we must take account of a large number for which the period and the major axis are statistically known. This is a fairly long task, and the results are summarized in table 7.9. The variation of h with spectral type arises from observational selection. It means that a pair of A-type or F-type stars is, on average, farther

Table 7.9 Values of h for Various Types of Class V.

Type								
A5	F0	F5	G0	G5	K0	K5	M0	M2
1.25	1.25	1.25	1.30	1.40	1.50	1.60	1.70	1.90

from periastron when it is discovered than is a pair of K-type or M-type stars. The standard deviation of these values is about ±0.13. This deviation arises from the somewhat arbitrary values of h. The period calculated in this way is a probable period. The true period has one chance in two of lying between $0.75P$ and $1.4P$.

Thanks to equation (7.17), we can estimate the rapidity of orbital motion as soon as a system is discovered. This gives us guidance as to how frequently we should observe the system, and so we can choose which pairs should be kept under annual surveillance. Also, the foregoing considerations allow us to guess the luminosity class, which gives the absolute magnitude, if the spectral type is known.

Table 7.10 is for main-sequence stars. It facilitates the calculation of probable periods. In it are to be found the spectral type, the absolute bolometric magnitude, the mass μ according to the mass-luminosity relation of Baize and Romani (chapter 6), and the term $(k/2 - 0.3)M$, which is used to calculate the probable period, with the value of k set at 0.1117.

Table 7.11 makes possible calculation of the magnitude of component A from the difference in magnitude between the components, Δm, and the combined visual magnitude.

Examples
• Consider a pair whose orbit is known: ADS 8325 = Hu 731 (1950 position 11^h49^m4 $+48°22'$; combined visual magnitude 8.6, $\Delta m = 0.2$, and spectral type K0, whence $M = 6.0$.) Its separation at discovery was $0''34$. We have $m_{vA} + C_{vA} = 9.1$. From table 7.10, $h = 1.50$, whence $P_{prob} = 117$ years. The actual period is 107 years.
• Consider now a pair whose orbit is unknown, but which shows appreciable orbital motion: ADS 1273 = A944 AB (1950 position 01^h35^m0 $+45°45'$; combined visual magnitude 8.2, and $\Delta m = 0.3$). Its proper motion is $0''015$ and its spectral type is A0, so it is probably of class V. We find $m_{vA} + C_{vA} = 8.4$, and $M_A = 0.3$. Its separation at discovery was $0''44$. The table gives $h = 1.25$, whence $P_{prob} = 1,500$ years. The pair has traversed 20° in 65 years.
• Finally, consider a pair discovered in 1971, namely Muller 224 (1950 position 19^h40^m7 $+76°18'$). Its combined visual magnitude is 8.0, and $\Delta m = 0.7$. Its proper motion is $0''20$ and spectral type is K0;

Table 7.10 The Main Sequence.

Spectrum	M	μ	(k/2 − 0.3)M
B7	−0.6	3.98	+0.15
B8	−0.3	3.68	+0.07
B9	0	3.41	0
A0	+0.3	3.16	−0.07
A1	0.9	2.71	−0.22
A2	1.3	2.50	−0.29
A3	1.7	2.15	−0.44
A4	2.0	2.04	−0.49
A5	2.2	1.94	−0.54
A6	2.4	1.84	−0.59
A7	2.6	1.75	−0.63
A8	2.8	1.66	−0.68
A9	2.9	1.62	−0.71
F0	3.1	1.58	−0.73
F1	3.2	1.50	−0.78
F2	3.3	1.46	−0.81
F3	3.4	1.42	−0.83
F4	3.5	1.39	−0.85
F5	3.6	1.35	−0.88
F6	3.8	1.28	−0.93
F7	3.9	1.25	−0.95
F8	4.1	1.19	−1.00
F9	4.2	1.16	−1.02
G0	4.4	1.10	−1.07
G1	4.5	1.07	−1.10
G2	4.7	1.02	−1.15
G3	4.8	0.99	−1.17
G4	5.0	0.94	−1.22
G5	5.1	0.92	−1.24
G6	5.3	0.87	−1.29
G7	5.5	0.83	−1.34
G8	5.6	0.81	−1.37
G9	5.8	0.77	−1.42
K0	6.0	0.73	−1.46
K1	6.2	0.69	−1.51

Table 7.10 (cont.)

Spectrum	M	μ	$(k/2 - 0.3)M$
K2	6.5	0.66	−1.56
K3	6.9	0.58	−1.68
K4	7.2	0.54	−1.76
K5	7.5	0.46	−1.83
K6	7.9	0.45	−1.93
K7	8.2	0.41	−2.00
K8	8.5	0.38	−2.07
K9	8.9	0.35	−2.17
M0	9.2	0.32	−2.24
M1	9.7	0.28	−2.37
M2	10.1	0.25	−2.46
M3	10.6	0.22	−2.59
M4	11.3	0.19	−2.76

Source: J.-C. Pecker and E. Schatzman, *Astrophysique générale* (Paris: Masson, 1959).

Table 7.11 Magnitude of the Primary Component.

Δm	d
0	0.7
0.2	0.6
0.4	0.6
0.6	0.5
0.8	0.4
1.0	0.3
1.2	0.2
1.5	0.2

therefore it is of class v. $M_A = 6.0$ and $m_{vA} + C_{vA} = 8.2$. Its separation is 0".17; $h = 1.50$, whence $P_{prob} = 23$ years.

Thanks to tables 7.10, 7.11, and 6.12 (which gives bolometric corrections) it is easy to apply equation (7.17) to any pair at any time, even as soon as it is discovered. Note, in the first place, that only very close pairs are likely to have short periods. It is among those, therefore, that we should look. The magnitude is of considerable importance, as important as the spectral type. Pairs of high temperature, at the head of the main sequence, have longer periods (other things being equal) than do red dwarfs.

Table 7.12 gives a selection of a few recently discovered pairs that are probably of short period. The luminosity class is not known, but the proper motions of all the pairs are such that they are certainly made up of ordinary dwarfs, except for COU 14, already known as a giant of class III. This list is evidence of the large number of discoveries still to be made. They will make an essential contribution to the calibration of the mass-luminosity relation.

CONTINUITY OF OBSERVATIONS: AN APPEAL TO MONASTERIES

The study of double stars can be adapted to the power of the instrument available. Certainly, the search for new pairs and the measurement of close binaries are reserved for large apertures. Many double stars, however, such as Castor, γ Virginis, or ζ Cancri, can be followed with average or even small instruments. The amateur will always find a choice, and his observations, if well made, will be useful. Amateurs need not necessarily be motivated by desire to do some scientific work, but rather by the love of observation. They work above all for themselves, as artists. Many amateur clubs and private observatories have instruments of small or average size. Nevertheless, only rarely do amateurs make measurements of double stars; they seem more interested in the photographic performance of their instruments, rather than the visual. The result is that measurements of double stars are almost all made by professionals—and only a few of them devote themselves to the task.

It is astonishing that in the whole world there are not more than

200 Voyage to the Country of Double Stars

Table 7.12 Recently Discovered Binaries Whose Periods Are Probably Short.

Name[a]	m_{vT}[b]	α 1950 δ	Spectral Type	ρ[c]	Year Discovered	P_{prob} (Years)	P_{obs}[d] (Years)
COU 854	8.3	0h58m6 + 35°19′	F8V	0″.17	1972	55	(60)
COU 1067	8.5	2 06 .0 + 35 26	F5V	0.15	1973	68	
COU 79	5.6	2 12 .9 + 24 49	F5V	0.25	1965	18	37
COU 1371	8.8	2 37 .8 + 38 52	F2V	0.17	1975	40	
COU 691	8.5	3 39 .1 + 31 31	F8V	0.14	1972	47	
COU 929	7.5	7 53 .1 + 23 50	G0V	0.18	1973	34	(30)
COU 169	9.6	10 11 .3 + 22 42	K5V	0.19	1967	36	85
COU 292	8.0	10 24 .2 + 19 46	F8V	0.19	1968	49	43
COU 1260	9.0	11 19 .4 + 37 21	G0V	0.18	1975	30	
WOR 24	9.4	13 29 .7 + 31 25	M0V	0.36	1960	29	**36**
COU 606	9.4	14 11 .6 + 31 14	M0V	0.18	1971	10	35
Mlr 347	8.8	15 32.4 + 84 41	K0V	0.17	1972	48	(45)
COU 798	8.6	15 32 .6 + 27 05	G0V	0.16	1972	65	33
COU 612	8.0	15 36 .9 + 25 54	G5V	0.15	1971	32	
COU 613	8.6	15 38 .6 + 31 38	G5V	0.27	1971	101	(108)
COU 1445	7.5	15 40 .2 + 42 13	F8V	0.22	1976	49	
COU 490	8.5	16 43 .0 + 29 33	F5V	0.15	1972	68	(72)
Mlr 182	8.4	16 45 .3 + 71 51	G5V	0.22	1971	65	(106)
COU 1289	7.8	16 56 .7 + 39 47	G0V	0.14	1975	30	
COU 1291	8.5	17 05 .8 + 38 14	G5V	0.14	1975	43	(10)
COU 1145	7.1	17 47 .2 + 37 05	G0V	0.15	1974	21	(27)

COU 1462	9.1	19 07.1 + 33 59	K6V	0.20	1976	21	
COU 321	7.9	19 15.8 + 20 06	A5V	0.16	1968	111	60
Mlr 224	8.0	19 40.7 + 76 18	K0V	0.17	1971	23	(36)
COU 14	5.3	21 47.8 + 17 04	F2III	0.36	1959	75	**31**
COU 542	8.5	22 54.6 + 24 25	K0V	0.20	1970	49	44
Mlr 4	7.2	23 38.7 + 45 58	F5V	0.12	1953	20	**20**

a. COU = P. Couteau; Mlr = P. Muller; WOR = C. E. Worley.
b. Combined visual magnitude (source: SAO catalogue).
c. Separation at time of discovery.
d. Deduced from arc traversed. Bold type indicates orbit already computed; parentheses indicate less certain values.

six or seven people who regularly measure binaries. To be sure, powerful instruments are needed, under favorable skies. These conditions can be found, however, in almost every country. France has always been in the vanguard, as was Russia in the nineteenth century; the United States of America, where the biggest refractors are to be found, has also been in the forefront. Africa has contributed much. The young, however, are more attracted to the varied problems of astrophysics, to the detriment of positional astronomy, for which one must love continuity and perseverance.

There is no automatic assurance that there will be new observers. Any observatory oriented towards the study of double stars can easily abandon this kind of research and turn to some other speciality, just because of the personality of an astronomer. If only one could entrust this work to certain observatories by special regulation! In England, several clergymen made a name for themselves in double-star astronomy, such as the Reverend T. E. Espin, who had a private observatory. In France, at the beginning of the century, there was Jonckheere's observatory at Hem, near Lille, where more than a thousand pairs were discovered with a 35-cm refractor. This installation has become the Laboratoire d'Astronomie of the University of Lille. Some national observatories have even entrusted important instruments to amateurs. This has been standard practice in the United States. Paul Baize, a professional physician, made more than 20,000 measures with the 30-cm and 38-cm refractors at Paris, until his retirement and return to his native Normandy. Nice Observatory is the only French establishment where astronomers devote themselves entirely to the observation of double stars. It is not certain that it will always continue that way, because scientific programs depend on committees sensitive to new fashions, loving changes and not always appreciating the "passé".

The study of double stars requires continual observations, followed for centuries. It exceeds the term of a human life, but not of a group that dedicates itself to some task and remains stable for a long time. The continuity of observations could then be assured. For this purpose, it is all to the good that the maintenance of a refractor, although important, is not very onerous, and that visual observation is free—it uses up nothing. There are groups of the kind needed, in monasteries for example. They provide wonderful situations where a good refractor could find a refuge and certain use in a dome financed by donors and constructed close to or within the establishment. There could be one or two astronomer monks, whose training would require some short visits to observatories. A small library and one or two reasonably priced pocket electronic calculators would complete the scientific equipment. Astronomy is a typically monastic activity: It provides food for meditation and strengthens spirituality.

BIBLIOGRAPHY

The bibliography for this chapter is contained in those of the preceding chapters, but it is useful to add the following:

Couteau, P. "État actuel sur la découverte des étoiles doubles visuelles." *Revue de l'association française pour l'avancement des sciences* 2 (1971): 242.

———. "La mesure des étoiles doubles visuelles." *Astrophysics and Space Science* 11 (1971): 7.

———. "Contribution à l'étude de dénombrement des étoiles doubles visuelles." *Journal des Observateurs* 43 (1960): 41.

———. "La Recherche d'étoiles doubles: Sa motivation." *Astronomy and Astrophysics* 13 (1971): 345.

Heintz. W. D. "A statistical study of binary stars." *Journal of the Royal Astronomical Society of Canada* 63 (1969): 275.

The 1,500 stars discovered by the author have been published in sixteen series:

1st series: *Journal des Observateurs* 49, no. 6 (1966): 220. COU 34–72.
2nd series: Ibid. 50, fasc. 1 (1967): 33. COU 73–145.
3rd series: Ibid. 51, fasc. 1 (1968): 31. COU 146–245.
4th series: *Astronomy and Astrophysics* supplement 1 (1970): 105. COU 246–345.
5th series: Ibid. 1 (1970): 419. COU 346–445.
6th series: Ibid. 5 (1972): 167. COU 446–545.
7th series: Ibid. 6 (1972): 177. COU 546–645.
8th series: Ibid. 6 (1972): 419. COU 646–745.
9th series: Ibid. 10 (1973): 273. COU 746–845.
10th series: Ibid. 12 (1973): 137. COU 846–945.
11th series: Ibid. 15 (1974): 253. COU 946–1045.
12th series: Ibid. 20 (1975): 379. COU 1046–1200.
13th series: Ibid. 24 (1976): 495. COU 1201–1350.

14th series: Ibid. 29 (1977): 249. COU 1351–1500.
15th series: Ibid. 35 (1978): 197. COU 1501–1650.
16th series: In press. COU 1651–1850.

The stars discovered by Muller, presently numbering 565, have been published in the following:

Muller, P. "Résultats préliminaires d'une recherche systématique d'étoiles doubles nouvelles entre +60° et la pôle Boréal." *Publ. Obs. Paris,* June 1973.

The first 407 couples can be found therein. The others have been published in the following *Information Circulars* of the Commission on Double Stars of the International Astronomical Union:

No. 60 (July 1973). Mlr 408–430.
No. 61 (November 1973). Mlr 431–479.
No. 62 (March 1974). Mlr 480–499.
No. 64 (January 1975). Mlr 500–522.
No. 67 (November 1975), Mlr 523–542.
No. 70 (November 1976), Mlr 543–565.

The 120 double stars recently discovered with the 65-cm refractor of Belgrade Observatory are reported in the following:

Popovic, G. M. The First General Catalogue of Double Star Observations Made in Belgrade 1951–1971. *Publications of the Observatory of Belgrade* no. 19, 1974.

8
CATALOGUE OF 744 DOUBLE STARS FOR INSTRUMENTS OF ALL SIZES

WHY MAKE A CATALOGUE?

The sight of a double star in a refractor never leaves you indifferent. Thousands of pairs of stars offer themselves to every instrument. The amateur has a free choice, as much to satisfy his curiosity and love of observing as to test his eyesight and the quality of his equipment. He still needs adequate catalogues, however, neither too heavy nor too thin, in which he can find objects for each night of the year, of a difficulty proportionate to his means. Such catalogues have been published in the past, notably in the United States. In France, there are lists of double stars, by constellation, in C. Flammarion's work *Les Étoiles et les Curiosités du Ciel* (C. Marpon and E. Flammarion, 1882). This work has become difficult to find, however, and is out of date. Its author gives a description of each pair visible with the instruments of the time, with an estimate of the colors made by comparing them with those of precious stones. Modern catalogues are less poetic, but the dry approach is not an obstacle to lyricism or dreaming. There used also to be lists of double stars, from Paul Baize's pen, in the Flammarion almanacs. Unfortunately, these works now appear in so condensed a form that double stars are excluded. I should also mention the series of articles published in the *Bulletin de la Société astronomique de France* under the title "Review of the Constellations." These articles have been brought together in a pamphlet and published by the Society. Recently, Paul Muller has published a list of double stars for amateurs in the same *Bulletin* (July 1976). There are 76 objects from among the most interesting in the sky included in the list, which is embellished with notes and judicious comments. Telescopes of 25-cm or 30-cm aperture are more and more widely distributed, however, and permit a much wider variety of observations. The users of these telescopes need some real catalogues from which they can choose what to observe.

The aim of this catalogue is twofold: to provide a complete list of the brightest pairs, so that they can be located easily, and to give precise separations so that focal lengths can be checked by photography. In fact, the amateur is more and more a photographer; if he wishes to do useful work he must calibrate his field. Widely separated double stars provide such a calibration to perfection.

DESCRIPTION OF THE CATALOGUE

There are 744 objects listed in order of increasing right ascension. From left to right the columns give the following.
- A running number. If the pair is particularly interesting, there is an asterisk in front of the number.
- The name of the pair, as a double star. The abbreviations are those used in the central card index in Washington; the principal ones are noted .ere:

A	R. G. Aitken	Hu	W. J. Hussey
BU	S. W. Burnham	I	R. T. A. Innes
COU	P. Couteau	KUI	G. P. Kuiper
Da	W. R. Dawes	Mlr	P. Muller
H	W. Herschel	STF	W. Struve
h	J. Herschel	STT	O. Struve
Ho.	G. W. Hough		

- The name of the star in its constellation.
- Coordinates for the equinox 1950, in tenths of a minute.
- Visual magnitudes of the components. With a few rare exceptions the primaries are brighter than magnitude 8, the secondaries than magnitude 10.
- Position angles to the nearest degree.
- Separations.
- The corresponding year. Sometimes these results are replaced by an ephemeris calculated from an orbit.

An "N" in the rightmost column indicates that there is an explanatory note at the end of the table. These notes give, whenever applicable, the first measures and ephemerides, so that the separations can be adjusted by extrapolation to the date of observation. Many of these pairs have shown only insignificant motion since their discovery. In particular, included in the catalogue are pairs selected by P. Muller in his list of "New fundamental distances of double stars" [*Journal des Observateurs* 32 (1949): 114]. We have also made such a list for very close pairs [*Astronomy and Astrophysics* 2 (1969): 136], but they are, for the most part, outside the range of amateurs' instruments. The separations have been determined with minute care, in order to permit a rigorous test of the calibration of the field. Sometimes the notes, like those of Flammarion, give descriptions of the colors, or indicate the minimum aperture desirable for a good view of the pair.

ADVICE TO OBSERVERS

Many stars in the list are visible to the naked eye, and setting on them presents no great difficulty. Amateurs do not always have an equatorial mount with graduated circles, however. In this case you need to be very familiar with your instrument and to take a few precautions. A star chart will be very useful; you can set on stars invisible to the naked eye with the help of alignments that can be recognized after you have used the coordinates to draw them on the charts. This is easier with a refractor. It is recommended to fix circles graduated in altitude and azimuth to its altazimuth mounting. To set on a star with these circles, note the time at which a bright star crosses the meridian; then determine the hour angle of the double star (always observed close to the meridian). On setting the altitude to $\delta + L$, where L is the complement with respect to 90° of your latitude, you should find the object sought by a short sweep with the instrument. The author knows whereof he speaks, because during his summer vacations he is once again only an amateur astronomer like the rest of you, equipped with a 10-cm refractor on an altazimuth mount and without a finder. Despite this, with a certain amount of patience, he can find double stars in the field of the eyepiece; some of them are far from being visible to the naked eye.

Almost all the objects in the list are accessible to a mirror of 40 cm, as are many others that we have not mentioned. If the well-equipped amateur wishes to go farther, he will need the documentation of an observatory, and he will then be able to contribute to the painstaking work of measuring double stars.

THE CATALOGUE

1	2	3	4	5	6	7	8	9
1	STT 514		$0^h02^m.0 + 41°49'$	6.9–9.5	169°	5″.32	1958	
2	STF 3056		02.1 + 33 59	7.4–7.4	146	0.68	1975	
3	STF 3057		02.3 + 58 15	7.2–9.3	298	3.62	1973	
4	STF 3062		03.5 + 58 09	6.9–8.0	287	1.42	1980 Orb.	
5	STF 2		06.5 + 79 26	6.3–6.6	24	0.66	1980 Orb.	
6	STT 1		09.2 + 65 51	7.2–9.9	208	1.38	1963	
7	BU 255		09.3 + 28 09	7.5–7.8	77	0.56	1975	
8	STF 13		13.4 + 76 40	6.6–7.1	57	0.86	1980 Orb.	
9	STF 19		14.1 + 36 20	7.0–9.5	139	2.26	1961	
10	STF 24		15.9 + 25 52	7.2–8.0	249	5.20	1967	
11	AC 1		18.3 + 32 42	7.5–8.0	290	1.76	1970	N
12	BU 1093		18.3 + 10 42	7.1–8.0	108	0.60	1973	N
13	STT 6		18.6 + 66 44	7.2–8.2	151	0.62	1980 Orb.	
14	Ho 210		23.0 + 36 12	8.0–9.7	75	0.95	1967	
*15	STT 12	λ Cas	29.0 + 54 15	5.6–5.9	183	0.58	1980 Orb.	
16	A 911		30.7 + 47 22	7.9–8.6	318	0.55	1963	
17	Se 1		31.9 – 4 49	7.5–8.0	258	0.54	1965	
18	STF 42		33.4 + 29 44	7.9–8.7	24	6.01	1974	
*19	STF 46	55 Psc	37.3 + 21 10	5.5–8.2	193	6.64		N
20	STF 48		39.5 + 71 06	7.0–7.2	334	5.42	1962	

Catalogue

1	2	3	4	5	6	7	8	9
21	STT 18		$0^h39^m.8 + 3°54'$	7.4–9.5	199°	1".47	1980 Orb.	
22	STF 52		41.4 + 45 58	8.0–9.0	12	1.42	1974	
23	STF 55		41.7 + 33 21	8.0–8.8	328	2.20	1963	
24	STF 59		45.2 + 51 10	7.2–8.1	144	2.13	1965	
*25	STF 60	η Cas	46.1 + 57 33	4.0–7.6	307	11.98	1980 Orb.	N
*26	STF 61	65 Psc	47.2 + 27 26	6.0–6.0	116	4.34	1967	
27	STF 65		49.6 + 68 36	8.0–8.0	38	3.14	1968	
28	STT 20	66 Psc	51.9 + 18 55	5.9–7.0	224	0.48	1980 Orb.	
29	Hu 802		52.0 + 49 08	7.2–7.8	213	0.37	1966	
*30	STF 73	36 And	52.3 + 23 22	6.2–6.8	262	0.65	1980 Orb.	
31	BU 302		55.6 + 21 08	6.7–8.1	153	0.43	1974	N
32	STF 79		57.2 + 44 27	6.0–7.0	193	7.87	1958	
33	Ma 1		57.7 + 47 03	7.9–8.4	13	0.95	1974	
34	A 2901		58.2 + 69 05	7.6–7.6	45	0.39	1962	N
35	BU 1161		1 00.0 + 51 32	6.9–7.7	355	0.38	1961	
36	STT 21		00.1 + 47 07	6.9–8.2	175	0.89	1980 Orb.	
37	BU 396		00.5 + 60 48	6.1–8.5	69	1.33	1965	
38	Hu 517		00.8 + 50 10	7.8–8.2	23	0.48	1971	
39	Ho 213		01.2 + 35 12	7.0–7.0	91	0.29	1974	N
*40	STF 88	ψ¹ Psc	03.0 + 21 13	4.9–5.0	160	29.97		

210 Catalogue

1	2		3		4	5	6	7	8	9
41	STF	90	77	Psc	1h03m.2 + 4°39′	5.9–6.4	83°	32″.91	1970	
42	STF	91			04.6 − 2 00	6.7–7.5	316	4.23	1970	
43	STT	515	φ	And	06.6 + 46 59	4.9–6.5	140	0.47	1980 Orb.	
44	BU	303			07.0 + 23 32	7.1–7.3	290	0.66	1967	
45	BU	235			07.6 + 50 45	7.0–7.4	124	1.04	1975	N
46	STF	96			09.4 + 64 45	7.8–8.8	286	0.96	1973	
47	BU	258			10.0 + 61 27	6.2–9.0	262	1.24	1973	
*48	BU	1029	ζ	Psc	11.1 + 7 19	4.2–5.3	63	23.20		N
49	BU	1100			11.6 + 60 41	7.4–7.4	37	0.53	1980 Orb.	
50	STT	28			14.3 + 80 36	7.0–8.5	295	0.78	1964	
51	STF	113	42	Cet	17.2 − 0 46	6.2–7.2	11	1.59	1974	N
52	BU	4			18.7 + 11 17	7.0–7.5	113	0.41	1980 Orb.	
53	STF	115			20.1 + 57 53	7.3–7.5	137	0.51	1974	
54	BU	1164	95	Psc	25.1 + 5 06	6.7–7.0	156	0.44	1980 Orb.	
55	STF	138			33.4 + 7 23	7.3–7.3	53	1.71	1975	
56	BU	5			36.6 + 16 22	7.0–9.0	288	0.78	1967	N
57	STF	141			37.1 + 38 43	8.0–8.5	304	1.60	1966	
58	Kr	12			38.1 + 62 28	7.7–7.7	119	0.41	1964	
*59	STF	147	χ¹	Cet	39.3 − 11 34	6.0–7.3	89	2.03	1962	N
60	BU	870			41.0 + 57 17	6.9–8.3	20	1.00	1965	

1	2		3		4	5	6	7	8	9
61	STF	155			1ʰ41ᵐ.6 + 9°14'	7.5–7.9	328°	4".83	1962	
62	BU	6			42.2 − 7 01	6.4–9.2	165	2.25	1974	
63	STT	34			44.3 + 80 38	7.3–7.5	280	0.44	1980 Orb.	
64	STF	162			46.2 + 47 39	7.0–7.5	206	2.01	1975	
65	STF	174	1	Ari	47.4 + 22 02	6.2–7.4	166	2.90	1965	
66	Ho	311			48.4 + 24 25	7.0–7.2	209	0.30	1973	
*67	STF	93	α	UMi	48.8 + 89 02	2.0–9.0	217	18.28		N
68	STF	178			49.4 + 10 34	7.8–7.8	200	3.10	1952	
69	STF	179			50.2 + 37 05	6.7–7.7	158	3.30	1966	
70	STF	170			50.6 + 75 59	6.7–7.5	248	3.30		
*71	STF	180	γ	Ari	50.8 + 19 03	4.2–4.4	0	7.82		N
72	STF	182			52.9 + 61 02	7.0–7.0	123	3.59	1953	
73	STF	186			53.3 + 1 36	7.2–7.2	53	1.22	1980 Orb.	
74	STF	194			56.5 + 24 35	8.0–8.3	276	1.09	1965	
75	BU	513	48	Cas	57.8 + 70 40	5.0–7.5	200	0.70	1980 Orb.	
*76	STF	202	α	Psc	59.4 + 2 31	4.0–5.0	288	1.95	1974	
*77	STF	205	γ	And	2 00.8 + 42 06	3.0–5.0	63	10.01		N
*78	STT	38	γ	And BC		5.0–6.2	108	0.55	1980 Orb.	N
79	BU	516			02.7 − 1 12	8.0–8.0	305	0.62	1963	
80	STF	212			03.4 + 24 25	8.0–8.5	163	1.92	1971	

212 Catalogue

1	2		3		4	5	6	7	8	9
81	STF	222	59	And	$2^h07^m.8 + 38°48'$	6.7–7.2	35°	16".64	1923	
82	STF	224			08.2 + 13 27	7.5–8.0	243	5.98	1959	
83	STF	226			09.4 + 23 44	7.8–9.7	238	1.82	1972	
*84	STF	227	ι	Tri	09.5 + 30 04	5.0–6.4	71	3.84	1970	
85	STF	231	66	Cet	10.2 − 2 38	6.0–7.8	235	16.17	1958	
86	STF	228			10.8 + 47 15	6.7–7.6	266	1.05	1980 Orb.	
87	STF	232			11.8 + 30 10	7.5–7.5	245	6.44	1965	
88	STF	249			18.4 + 44 22	7.0–9.0	195	2.19	1969	
*89	STF	262	ι	Cas	24.9 + 67 11	4.2–7.1	236	2.34	1980 Orb.	N
90	KUI	8			25.4 + 1 44	6.1–6.5	32	0.50	1970	N
91	STF	268			25.9 + 55 19	6.9–8.2	131	2.70	1963	
92	STF	274			28.9 + 0 52	7.2–7.7	220	13.82	1958	
93	STF	280			31.6 − 5 51	7.5–7.7	346	3.58	1961	
94	STF	281	ν	Cet	33.2 + 5 23	5.9–9.6	83	7.78		
95	STF	5	30	Ari	34.1 + 24 26	6.1–7.1	274	38.18	1953	
96	STF	285			35.9 + 33 12	7.0–7.7	167	1.74	1963	
97	STF	283			36.7 + 61 16	8.0–8.8	208	1.73	1969	
98	STF	289	33	Ari	37.8 + 26 51	5.8–8.7	0	28.80		
99	STT	43			37.8 + 26 25	7.2–8.8	8	1.03	1980 Orb.	
100	STF	291			38.3 + 18 35	7.4–7.7	116	3.40	1968	

213 Catalogue

1	2		3		4	5	6	7	8	9
101	STT	45			2^{h}38^m.3 + 4°39'	7.0–9.2	275°	1".05	1973	
102	STF	295	84	Cet	38.6 − 0 54	6.0–9.2	312	4.09	1968	
103	STT	44			39.0 + 42 29	7.8–8.5	58	1.40	1969	
*104	STF	299	γ	Cet	40.7 + 3 02	3.0–6.8	297	2.79	1966	
105	STF	300			41.6 + 29 15	7.9–8.1	311	3.13	1963	
106	BU	9			44.0 + 35 21	6.3–8.4	191	1.52	1975	
107	STF	305			44.6 + 19 10	7.3–8.2	309	3.64	1980 Orb.	
*108	STF	311	π	Ari	46.5 + 17 15	4.9–8.4	118	3.17	1969	N
109	STF	315			46.9 − 10 45	7.5–8.7	162	1.83	1964	
*110	STF	307	η	Per	47.0 + 55 41	4.0–8.5	301	28.42		
111	STF	306			47.2 + 60 13	7.1–9.0	95	2.09	1965	
112	STF	314			49.3 + 52 48	6.9–7.1	307	1.45		N
113	STF	312			51.0 + 72 41	7.1–8.0	35	2.33	1974	N
114	BU	525			56.0 + 21 25	7.2–7.2	267	0.42	1980 Orb.	
*115	STF	333	ε	Ari	56.4 + 21 08	5.7–6.0	207	1.36	1972	N
116	STF	334			56.7 + 6 27	7.7–8.2	312	1.18	1972	
*117	STF	331			57.2 + 52 09	5.3–6.7	86	12.18		N
118	BU	11	ρ²	Eri	3 00.2 − 7 53	5.4–9.6	71	1.79		
119	STF	349			06.6 + 63 36	7.4–8.1	321	6.10	1962	
120	STT	50			07.6 + 71 02	7.5–7.5	170	1.15	1975	

214 Catalogue

1	2	3		4	5	6	7	8	9
121	STF 360			3ʰ09ᵐ.0 + 37°02'	7.8–8.0	127°	2″.72	1974	N
122	STT 51			09.6 + 44 06	7.9–8.1	324	1.20	1975	
*123	h 3555	α	For	09.9 − 29 12	4.0–7.0	305	3.00	1980 Orb.	
124	STF 367			11.5 + 0 33	8.0–8.0	143	1.00	1980 Orb.	
125	STT 52			13.1 + 65 29	6.4–7.0	69	0.45	1980 Orb.	
126	BU 84			13.6 − 6.06	7.2–7.4	15	0.85	1971	
127	STF 369			13.9 + 40 18	6.5–7.8	30	3.56	1958	
128	STT 53			14.5 + 38 27	7.2–8.0	265	0.86	1980 Orb.	
129	COU 259	τ	Ari	18.3 + 20 58	5.0–8.4	246	0.66	1970	N
130	Hu 1058			21.7 + 40 03	7.8–8.5	116	0.80	1967	
131	STF 384			24.4 + 59 44	7.8–9.0	269	2.02	1966	
132	STF 394			25.0 + 20 17	7.0–8.0	163	6.80	1968	N
133	STF 391			25.8 + 44 53	7.3–8.0	95	3.69	1923	
134	STF 389			26.1 + 59 12	7.0–8.0	70	2.62	1964	
135	STF 401			28.3 + 27 24	6.5–7.0	270	1.40	1967	
136	STF 400			30.9 + 59 52	7.0–8.0	258	1.25	1980 Orb.	
137	STF 412	7	Tau	31.5 + 24 18	6.6–6.7	6	0.65	1980 Orb.	
138	BU 533			32.5 + 31 31	7.0–7.0	224	1.05	1973	
139	STF 422			34.2 + 0 26	6.0–8.2	264	6.60	1967	
140	STF 425			37.0 + 33 57	7.3–7.3	78	1.84	1968	N

215 Catalogue

1	2		3		4	5	6	7	8	9
141	STT	59			$3^h37^m.2 + 45°52'$	7.5–7.8	354°	2".57	1973	
142	STF	419			37.6 + 69 41	7.2–7.2	73	2.78	1967	N
143	STF	427			37.6 + 28 37	6.6–7.4	209	6.86		
144	STF	431	40	Per	39.2 + 33 48	5.0–9.5	238	20.0		
145	BU	535	o	Per	41.2 + 32 08	4.0–8.5	30	1.03	1963	
146	Ho	504			41.4 + 35 42	7.8–8.0	195	1.14	1968	
147	STT	516			46.0 + 32 07	7.2–9.2	43	2.27	1967	
*148	STT	65			47.3 + 25 26	6.5–6.8	206	0.71	1980 Orb.	
149	STT	66			48.7 + 40 39	7.5–8.0	144	0.89	1964	
*150	KUI	15	31	Tau	49.3 + 6 23	5.5–5.6	211	0.55	1967	N
*151	STF	464	ζ	Per	51.0 + 31 44	2.7–9.3	208	12.9		
*152	STF	470	W	Eri	51.8 − 3 06	4.0–6.0	345	6.90	1963	
153	Hn	67			52.2 − 12 52	7.8–8.5	155	2.75	1951	
154	STT	67			52.9 + 60 58	5.0–8.2	46	1.78	1963	
155	STF	471	ε	Per	54.5 + 39 52	3.1–8.3	9	8.99		
156	STT	69			56.3 + 38 41	6.4–9.1	326	1.87		
157	STF	479			58.0 + 23 04	7.0–7.9	127	7.47	1957	
*158	STF	460	49	Cep	4 01.4 + 80 34	5.2–6.1	112	0.77	1980 Orb.	
159	STF	485			03.4 + 62 12	6.1–6.2	305	17.92	1954	N
160	STT	531			04.2 + 37 57	6.5–8.2	3	1.45	1980 Orb.	

216 Catalogue

1	2		3		4	5	6	7	8	9
161	STF	495			4h04m.8 + 15°02'	6.0–8.8	222°	3".80	1972	
162	STT	72			05.1 + 17 12	6.1–9.2	326	4.44		
163	STF	494			05.9 + 22 58	7.7–7.7	187	5.25		
164	BU	546			08.0 + 41 44	8.0–8.0	39	0.98	1966	
165	BU	547	47	Tau	11.2 + 9 08	5.5–8.0	347	1.25	1970	
166	STT	77			12.8 + 31 34	7.5–7.5	271	0.78	1980 Orb.	
167	STF	511			13.7 + 58 40	7.5–8.0	107	0.40	1980 Orb.	
168	STT	75			14.3 + 60 22	7.6–8.0	174	0.46	1969	
169	BU	87			19.4 + 20 42	5.7–8.8	169	2.02	1970	
170	STF	528	χ	Tau	19.5 + 25 31	5.7–7.8	24	19.42	1955	
171	STT	82			19.9 + 14 56	7.0–9.0	359	1.40	1980 Orb.	
172	STT	80			20.1 + 42 19	6.5–7.0	166	0.44	1969	
173	STF	535			20.5 + 11 16	6.7–8.2	302	1.36	1970	
174	Hu	304	66	Tau	21.1 + 9 21	5.9–5.9	71	0.21	1980 Orb.	
175	Ho	15			21.3 + 30 01	8.0–8.0	145	0.72	1968	
*176	KUI	17	δ	Tau	22.6 + 17 48	4.2–7.9	334	1.54	1968	N
177	STF	531			22.7 + 55 32	7.4–8.6	317	1.10	1966	
178	BU	311			24.8 − 24 11	6.5–7.0	116	0.53	1980 Orb.	
179	BU	184			25.8 − 21 37	6.2–7.0	256	1.48		
180	STF	554	80	Tau	27.3 + 15 32	6.5–9.0	16	1.81	1980 Orb.	

217 Catalogue

1	2		3		4	5	6	7	8	9
181	STF	552			4ʰ28ᵐ.0 + 39°54'	6.3–6.5	115°	9".1		
*182	STF	550	1	Cam	28.1 + 53 48	5.1–6.2	308	10.2	1964	
183	STF	559			30.6 + 17 55	7.0–7.1	276	2.90	1972	
184	STT	86			33.6 + 19 40	7.5–7.5	25	0.46	1972	
185	STF	565			34.6 + 42 01	7.2–8.5	170	1.37	1968	
*186	STF	572			35.4 + 26 51	6.5–6.5	194	4.03	1961	
187	KUI	18	1	Eri	35.9 – 14 24	4.0–7.0	52	0.73	1961	N
188	STF	566	2	Cam	36.0 + 53 23	5.4–7.4	229	0.81	1980 Orb.	N
189	STF	577			38.8 + 37 25	7.7–7.7	25	1.01	1975	N
190	STF	590	55	Eri	41.2 – 8 53	6.6–6.8	318	9.28	1953	
191	STT	90			52.2 + 8 31	7.0–9.0	340	1.94	1969	
192	STT	88			52.7 + 61 41	6.5–8.2	310	0.79	1974	
193	STT	91			53.6 + 3 06	7.0–7.5	231	0.53	1968	
194	STT	92	5	Aur	56.9 + 39 19	6.0–9.7	272	3.70	1963	N
195	STF	631			58.4 – 13 34	7.2–8.7	106	5.68		
196	STF	636			5 00.6 – 8 44	7.5–8.6	104	3.70	1971	
197	STT	95			02.6 + 19 44	6.6–7.2	307	0.90	1968	
198	STT	98	14	Ori	05.2 + 8 26	6.0–6.8	21	0.68	1980 Orb.	N
199	STF	645			06.6 + 27 58	6.2–8.7	27	11.60	1961	
200	STF	644			06.9 + 37 14	6.7–7.0	221	1.63	1960	

218 Catalogue

1	2	3	4	5	6	7	8	9
201	STF 648		$5^h07^m.8 + 31°59'$	7.4–8.1	66°	4".77	1957	
202	STF 652		09.2 + 0 59	6.3–7.8	182	1.67	1957	
*203	STF 654	ρ Ori	10.7 + 2 48	4.7–8.5	64	6.92	1961	N
204	STT 517		10.9 + 1 55	6.5–6.7	228	0.54	1980 Orb.	
*205	STF 661	κ Lep	10.9 – 13 00	5.0–7.9	359	2.64	1938	N
*206	STF 668	β Ori	12.1 – 8 15	1.0–8.0	203	9.50	1962	
207	STF 664		12.4 + 8 23	7.5–8.0	171	4.98		
208	STF 634	19 Cam	14.3 + 79 12	4.5–9.0	114	15.9	1961	N
209	STF 657		14.8 + 52 47	7.5–8.0	301	1.00	1974	
210	h 3750	38 Lep	18.3 – 21 17	4.8–9.5	281	4.36		
211	h 3752	41 Lep	19.8 – 24 49	5.5–6.7	99	3.14		
212	STF 677		20.0 + 63 21	7.7–8.0	158	1.04	1980 Orb.	N
213	Wnc 2		21.3 – 0 55	6.5–7.5	161	2.41	1952	
*214	Da 5	η Ori	22.0 – 2 26	4.0–5.0	81	1.50	1971	
215	Ho 226		23.9 + 27 34	7.0–7.0	257	0.81	1969	
*216	STF 716	118 Tau	26.2 + 25 07	5.8–6.6	206	4.83	1957	N
217	Da 6		26.5 – 3 20	7.2–7.5	143	0.27	1966	
*218	STF 728	32 Ori	28.1 + 5 55	5.2–6.7	44	0.93	1980 Orb.	N
*219	STF 729	33 Ori	28.6 + 3 15	6.0–7.3	26	1.88	1957	
220	BU 1049		30.6 – 1 45	7.0–8.6	355	1.46	1967	N

219 Catalogue

1	2	3	4	5	6	7	8	9
221	Mlr 314		$5^h32^m1 + 66°40'$	6.5–7.3	306°	0".27	1972	N
*222	STF 738	λ Ori	32.4 + 9 54	4.0–6.0	44	4.28	1955	
*223	STF 748	θ Ori	32.8 − 5 25	4.7		Multiple system		N
224	Da 4	42 Ori	32.9 − 4 52	5.5–8.7	207	1.49	1968	
225	STF 750		33.0 − 4 24	6.0–8.0	58	4.36	1940	
*226	STF 752	ι Ori	33.0 − 5 56	3.2–7.3	141	11.4		
227	STF 742		33.4 + 21 58	7.2–7.8	269	4.00	1968	N
228	STF 749		34.0 + 26 54	7.1–7.2	330	1.12	1966	N
229	STF 753	26 Aur	35.4 + 30 28	5.0–8.0	267	12.3	1960	
230	STT 112		36.5 + 37 56	7.3–8.0	55	0.72	1974	
*231	STF 774	ζ Ori	38.2 − 1 58	2.0–5.7	160	2.36		
232	STF 785		42.8 + 25 54	6.7–7.7	348	14.2		
233	BU 560		44.2 + 29 39	8.0–8.0	136	1.40	1971	
234	STF 3115		44.4 + 62 48	6.7–7.8	358	0.90	1969	
*235	STF 795	52 Ori	45.3 + 6 26	6.2–6.2	211	1.35		
236	STF 780		46.0 + 65 44	6.7–7.9	104	3.81		
237	SFF 813		51.4 + 18 53	8.0–8.0	148	3.08		
*238	STT 545	θ Aur	56.3 + 37 13	3.0–7.5	317	3.48	1967	N
239	STF 848		6 05.7 + 13 59	7.3–8.0	110	2.41	1964	
240	STF 855		06.4 + 2 31	5.8–6.8	113	29.4		

220 Catalogue

1	2		3		4	5	6	7	8	9
*241	STF	845	41	Aur	6^{h}07^m.8 + 48°44'	5.2–6.4	356°	7".66		
242	STF	867			08.7 + 17 23	7.0–8.5	158	2.22		N
243	KUI	24			11.6 + 17 55	6.5–6.5	320	0.56	1967	N
*244	BU	1008	η	Gem	11.9 + 22 31	3.0–6	265	1.40	1968	
245	STF	872			12.3 + 36 10	6.0–7.0	217	11.2		
246	STF	880			12.7 + 10 36	8.0–8.0	54	5.48		
247	STF	881	4	Lyn	17.6 + 59 24	6.4–7.9	129	0.78		N
248	A	2667			18.8 + 2 18	7.5–7.8	153	0.40	1980 Orb.	
249	STF	899			19.9 + 17 36	7.0–8.0	19	2.26		
250	STF	900	8	Mon	21.1 + 4 37	4.0–6.7	30	13.3		
*251	STF	919	β	Mon	26.4 – 7 00	5.0–5.5	132	7.25		N
252	STF	924	20	Gem	29.4 + 17 49	6.0–6.9	210	20.0		
253	STF	918			30.0 + 52 30	6.7–7.7	330	4.52		
254	STF	928			31.2 + 38 35	7.4–8.0	133	3.41	1955	
255	STF	941			35.2 + 41 38	7.0–8.0	82	1.77	1966	
256	STF	936			35.4 + 58 09	7.0–8.7	277	1.30	1967	
257	STT	152	54	Aur	36.4 + 28 19	6.0–7.8	34	0.96	1966	
258	STF	945			36.8 + 41 01	7.1–8.0	307	0.47	1974	N
259	Mlr	318			37.4 + 66 15	7.1–8.7	309	1.27	1972	N
260	STF	950	15	Mon	38.2 + 9 57	6.0–8.8	212	2.92	1963	

221 Catalogue

1	2	3	4	5	6	7	8	9
261	STF 946		$6^h40^m.4 + 59°30'$	7.2–9.0	130°	4".15	1963	N
262	STT 154		40.8 + 40 41	6.7–8.4	103	23.18	1963	N
*263	STF 948	12 Lyn	41.8 + 59 30	5.2–6.1	259	1.69	1980 Orb.	N
*264	Sirius		43.0 – 16 39	–1.6–8	49	10.25	1980 Orb.	
*265	STF 958		44.0 + 55 46	6.0–6.0	78	4.78	1962	
266	STT 156		44.5 + 18 15	6.5–7.0	243	0.54	1980 Orb.	
267	STT 157		45.2 + 0 24	7.5–8.0	218	0.34	1980 Orb.	
268	STF 963	14 Lyn	48.7 + 59 31	5.9–7.1	251	0.43	1980 Orb.	
*269	STF 982	38 Gem	51.8 + 13 15	5.4–7.7	151	6.98	1962	N
270	STF 981		52.2 + 30 14	8.0–8.0	136	1.68	1973	N
*271	STT 159	15 Lyn	53.0 + 58 29	5.1–6.2	33	0.97	1958	
*272	STF 997	μ CMa	53.8 – 13 59	4.7–8.0	339	3.00		
273	CO 7	ε CMa	56.7 – 28 54	1.6–8.0	160	7.45		
274	STF 973		57.5 + 75 19	6.6–7.6	30	12.49	1961	
275	STF 1009		7 01.7 + 52 50	6.7–6.8	149	3.93	1957	N
276	STF 1025		08.8 + 55 53	7.5–7.8	135	24.6		
277	STF 1037		09.7 + 27 19	7.1–7.1	324	1.13	1973	N
278	STF 1033		10.8 + 52 38	7.4–8.0	278	1.49		
279	STF 1056		13.0 – 1 46	7.8–8.8	299	3.85		
280	h 3945	145 CMa	14.5 – 23 13	5.0–7.0	58	27.4		

222 Catalogue

1	2	3		4	5	6	7	8	9
281	STT 170			7ʰ14ᵐ9 + 9°23'	7.5–7.8	92°	1″.38	1966	
*282	STF 1066	δ	Gem	17.1 + 22 05	3.2–8.2	217	6.33	1961	N
283	STF 1074			18.0 + 0 30	7.8–8.2	163	0.62	1965	N
284	STF 1065	20	Lyn	18.4 + 50 15	6.6–6.8	254	15.1		
285	STF 1062	19	Lyn	18.8 + 55 23	5.3–6.6	315	14.7		
286	STF 1081			21.2 + 21 33	7.8–8.5	230	1.80	1972	
287	STF 1083			22.6 + 20 36	6.7–7.8	45	6.71	1956	
288	STF 1094			24.6 + 15 25	7.7–8.7	96	2.44	1971	
289	STF 1104			27.1 − 14 53	6.7–8.3	3	2.14	1965	N
290	STF 1103			27.9 + 5 22	7.0–8.5	245	4.16	1965	
*291	STF 1110		Castor	31.4 + 32 00	2.7–3.7	98	2.31	1980 Orb.	N
292	STF 1116			31.8 + 12 25	7.0–7.7	100	1.78	1969	
293	STT 174			32.4 + 43 09	6.5–8.1	86	2.08	1967	
294	STF 1121			34.3 − 14 22	7.2–7.5	304	7.44		
295	STF 1126			37.5 + 5 21	7.2–7.5	160	1.01	1966	N
*296	STT 179	κ	Gem	41.4 + 24 31	4.0–8.5	241	6.47	1960	
297	STF 1127			42.4 + 64 11	6.2–8.0	340	5.40		
298	STF 1138	2	Pup	43.2 − 14 34	6.2–7.0	339	16.90		
299	STF 1140			45.5 + 18 28	6.8–8.5	274	6.22		
300	Wil 15	82	Gem	45.6 + 23 16	6.0–6.0	49	0.24	1975	N

223 Catalogue

1	2	3		4	5	6	7	8	9
301	STF 1146	5	Pup	7^{h}45^m.6 − 12°04′	5.3–7.4	358°	1″.88	1974	N
302	BU 1195			48.9 − 9 17	7.3–7.6	87	0.34	1950	
303	BU 101	9	Arg	49.5 − 13 46	5.6–6.2	44	0.19	1980 Orb.	N
304	STT 182			50.1 + 3 31	7.0–7.5	20	1.04	1966	
305	STF 1157			52.0 − 2 40	8.0–8.0	212	0.90	1974	N
306	STF 1175			59.8 + 4 18	7.8–9.7	262	1.26	1973	N
307	STT 186			8 00.3 + 26 25	7.5–8.2	73	0.94	1966	
308	STT 187			01.0 + 33 10	6.9–7.5	355	0.31	1980 Orb.	
309	BU 581			01.6 + 12 26	8.0–8.0	254	0.48	1980 Orb.	
*310	STF 1177			02.6 + 27 40	6.5–7.4	350	3.51	1968	
311	STF 1187			06.4 + 32 22	7.1–8.0	30	2.71	1968	N
*312	STF 1196	ζ	Cnc	09.3 + 17 48	5.0–5.5	81	5.74	1974	N
*313	STF 1196			id	5.0–5.7	280	0.81	1980 Orb.	N
314	STF 1216			18.8 − 1 26	7.5–8.2	275	0.61	1980 Orb.	
*315	STF 1224	24	Cnc	23.7 + 24 42	6.0–7.1	48	5.74	1968	N
*316	STF 1223	φ²	Cnc	23.8 + 27 06	6.0–6.5	218	5.17	1968	
317	− 489	9	Pyx	29.3 − 19 24	5.9–6.5	40	0.22	1972	
318	BU 205			30.9 − 24 26	7.0–7.0	2	0.56	1980 Orb.	
319	STF 1245			33.2 + 6 48	6.0–7.0	26	10.22		
320	STF 1260			38.3 − 12 00	7.8–8.3	302	5.15	1962	

224 Catalogue

1	2	3		4	5	6	7	8	9
321	Ho 529			$8^h39^m.8 - 17°15'$	7.6–7.7	184°	0".35	1968	
322	Ho 355			40.4 − 2 31	8.0–8.0	162	0.70	1967	N
323	STF 1270			42.8 − 2 25	6.6–7.6	261	4.66		
324	STF 1268	ι	Cnc	43.7 + 28 57	4.4–6.5	307	30.4	1958	N
*325	STF 1273	ε	Hya	44.2 + 6 36	3.8–7.8	275	2.84	1968	N
326	STF 1275			47.6 + 57 43	8.0–8.0	196	1.81	1966	
*327	STF 1282			47.6 + 35 15	7.0–7.0	279	3.60	1968	
328	A 2473			47.8 + 18 11	7.7–7.7	27	0.32	1971	
329	STF 1280			51.0 + 71 00	7.5–7.6	123	1.20	1980 Orb.	
*330	STF 1291	57	Cnc	51.2 + 30 46	5.9–6.4	317	1.46	1968	
331	STF 1289			51.4 + 43 47	7.7–8.5	5	3.71	1967	
332	STF 1295	17	Hya	53.0 − 7 47	7.2–7.3	0	4.30		
333	Arg 72			54.5 − 17 14	7.7–7.9	2	3.46	1962	
*334	STT 196	ι	UMa	55.8 + 48 14	3.1–9.0	21	4.37	1969	N
335	KUI 37	10	UMa	57.6 + 42 00	4.3–6.3	27	0.69	1980 Orb.	N
336	STF 1298			58.4 + 32 27	6.1–8.2	136	4.60		
*337	A 1585	κ	UMa	$9^h00.2 + 47 21$	4.0–4.2	284	0.26	1980 Orb.	N
338	STF 1311			04.6 + 23 11	6.7–7.1	200	7.60	1968	
339	STF 1306	σ^2	UMa	06.0 + 67 20	5.0–8.2	3	3.22	1980 Orb.	
340	STF 1312			06.8 + 52 35	7.7–8.2	148	4.46		

225 Catalogue

1	2	3	4	5	6	7	8	9
341	STF 1322		$9^h09^m9 + 16°44'$	7.7–8.2	52°	1″71	1966	
342	STF 1318		10.3 + 47 12	7.5–8.7	234	2.58	1973	N
343	BU 212		13.6 − 8 08	7.5–8.2	207	1.40	1963	
344	STF 1332		14.4 + 23 52	7.2–7.5	26	5.81	1962	
345	STF 3121		14.9 + 28 47	7.5–7.5	170	0.17	1980 Orb.	N
*346	STF 1333		15.4 + 35 35	6.6–6.9	48	1.74		
*347	STF 1334	38 Lyn	15.8 + 37 01	4.0–6.7	229	2.80		
348	STF 1331		16.9 + 61 34	8.0–8.0	154	0.91	1969	
349	STF 1338		17.9 + 38 24	7.0–7.2	256	0.92	1980 Orb.	
350	STF 1340	39 Lyn	19.2 + 49 46	6.5–8.3	319	6.19		
351	STF 1348		21.8 + 6 34	7.5–7.6	317	1.99	1966	
352	STF 1346		22.1 + 54 15	7.0–8.0	313	5.76	1957	
353	STF 1355		24.7 + 6 27	7.2–7.2	343	2.20	1965	
354	A 1588		24.8 − 9 00	7.2–7.2	197	0.27	1964	
*355	STF 1356	ω Leo	25.8 + 9 17	6.2–7.0	14	0.47	1980 Orb.	
356	A 1985		26.8 + 42 29	8.0–8.0	28	1.33	1970	N
357	STF 1351	23 UMa	27.6 + 63 17	3.8–9.0	271	22.8		
358	STF 1365		29.0 + 1 41	7.0–8.0	157	3.14	1974	
359	STF 1362		33.2 + 73 18	7.0–7.0	127	4.71	1969	
360	STF 1374		38.3 + 39 11	7.0–8.3	299	2.82	1969	

226 Catalogue

1	2		3		4		5	6	7	8	9
361	KUI	44	20	Leo	9h47m.0	+ 21°27'	6.6–6.9	216°	0".33	1971	N
362	Ho	369			48.2	+ 36 43	7.7–7.8	100	0.36	1973	N
*363	STT	208	φ	UMa	48.7	+ 54 18	5.0–5.6	100	0.24	1980 Orb.	
364	AC	5	8	Sex	50.0	− 7 52	5.5–5.9	79	0.50	1980 Orb.	
365	STT	210			59.4	+ 46 36	7.5–8.3	259	1.10	1966	
366	STT	215			10 13.6	+ 17 59	7.0–7.2	181	1.38	1980 Orb.	N
*367	STF	1415			13.9	+ 71 19	6.1–7.0	167	16.54	1958	
368	STF	1421			15.3	+ 27 47	7.5–8.5	330	4.44	1966	
*369	STF	1424	γ	Leo	17.2	+ 20 06	2.0–3.5	122	4.33	1968	N
370	STF	1426			17.9	+ 6 41	7.8–8.3	302	0.85	1968	
*371	Hu	879	β	LMi	25.0	+ 36 58	4.0–6.5	227	0.50	1980 Orb.	N
372	STF	1450	49	Leo	32.4	+ 8 55	6.0–8.7	158	2.20	1962	
373	STF	1457			36.1	+ 6 00	7.4–8.4	331	1.79	1967	N
374	STF	1466	35	Sex	40.8	+ 5 01	6.1–7.2	240	6.69	1974	
375	STT	229			45.2	+ 41 22	6.7–7.1	288	0.83	1968	N
376	STF	1476	40	Sex	46.8	− 3 46	7.2–8.0	10	2.31	1961	
*377	STF	1487	54	Leo	52.9	+ 25 01	5.0–7.0	110	6.44	1968	
378	STF	1504			11 01.4	+ 3 55	7.5–7.6	117	1.25	1973	
379	STF	1521			12.7	+ 27 51	7.2–7.5	96	3.59	1971	
*380	STF	1523	ξ	UMa	15.6	+ 31 49	4.0–4.9	105	2.92	1980 Orb.	

227 Catalogue

1	2	3		4	5	6	7	8	9
381	STF 1524	ν	UMa	11ʰ15.ᵐ8 + 33°22'	3.7–9.9	148°	7".21	1959	
*382	STF 1536	ι	Leo	21.3 + 10 48	3.9–7.1	154	1.22	1980 Orb.	N
383	STF 1537			21.8 + 20 54	7.6–8.6	357	2.39	1969	
384	h 840	γ	Crt	22.4 – 17 25	4.0–9.5	93	5.26	1968	
*385	STF 1540	83	Leo	24.3 + 3 17	6.3–7.3	150	28.70	1959	N
*386	STF 1543	57	UMa	26.4 + 39 37	5.2–8.2	358	5.34	1962	
387	STT 234			28.1 + 41 34	7.0–7.4	116	0.26	1980 Orb.	N
388	STT 235			29.5 + 61 22	6.0–7.3	202	0.36	1980 Orb.	N
389	HlII 96	17	Crt	29.8 – 28 59	6.0–6.0	210	9.10		
390	STF 1552	90	Leo	32.1 + 17 04	6.0–7.3	207	3.39	1968	
391	STF 1555			33.7 + 28 03	6.4–6.8	140	0.50	1974	
392	STF 1561			36.2 + 45 23	5.9–8.0	253	9.74	1958	
393	STT 237			36.3 + 41 25	7.4–9.0	250	1.62	1971	N
394	STT 241			53.7 + 35 44	6.5–8.4	141	1.75	1966	
*395	STF 1596	2	Com	12 01.7 + 21 44	6.0–7.5	237	3.78	1956	
396	STF 1606			08.3 + 40 10	6.3–7.0	264	0.38	1980 Orb.	
397	BU 920			13.2 – 23 04	6.5–7.0	278	1.17	1960	
398	STF 1622	2	CVn	13.6 + 40 56	5.7–8.0	260	11.60	1958	
399	STF 1625			14.1 + 80 25	6.5–7.0	219	14.4		
*400	STF 1627			15.6 – 3 40	5.9–6.4	196	20.05	1951	

1	2	3	4	5	6	7	8	9
401	STF 1639		$12^h21^m.9 + 25°52'$	6.7–7.9	326°	1″.49	1980 Orb.	
402	STT 250		22.0 + 43 22	7.7–8.0	343	0.45	1965	
*403	Sh 145	δ Cor	27.3 − 16 15	3.0–7.5	212	24.2		N
404	STF 1647		28.0 + 10 00	7.5–7.8	237	1.39	1966	
*405	STF 1657	24 Com	32.6 + 18 39	4.7–6.2	271	20.3	1958	
406	STF 1669		38.7 − 12 44	6.5–6.5	309	5.50	1957	
*407	STF 1670	γ Vir	39.1 − 1 11	3.0–3.0	297	3.92	1980 Orb.	N
408	STF 1678		42.9 + 14 39	6.3–7.0	181	34.3	1958	
*409	STF 1694		48.6 + 83 41	4.9–5.4	326	21.50	1958	
410	STF 1685		49.4 + 19 27	6.8–7.3	202	16.07		
*411	STF 1687	35 Com	50.8 + 21 31	5.0–7.8	163	1.06	1980 Orb.	N
*412	STF 1692	α CVn	53.7 + 38 35	3.2–5.7	228	19.44	1974	N
413	STT 256		53.9 − 0 41	7.2–7.6	89	0.92	1966	
414	STF 1695		54.1 + 54 22	6.3–8.2	282	3.74	1955	
415	STF 1699		56.3 + 27 45	7.8–7.8	7	1.64	1963	
416	BU 928		13 00.8 − 6 10	7.8–8.7	317	2.28	1965	
417	BU 929	48 Vir	01.3 − 3 24	6.2–6.5	205	0.75	1966	
418	COU 11	39 Com	03.9 + 21 26	6.0–9.0	322	1.31	1975	N
419	STF 1719		04.8 + 0 51	7.3–7.8	0	7.12		
420	STF 1722		06.0 + 16 46	7.8–8.8	337	2.80		

229 Catalogue

1	2	3		4	5	6	7	8	9
*421	STF 1724	θ	Vir	13h07m.4 − 5°16'	4.0–9.0	342°	7".10		
422	STF 1728	42	Com	07.6 + 17 47	6.0–6.0	191	0.49	1980 Orb.	N
423	STT 261			09.7 + 32 21	6.9–7.4	340	2.24	1966	N
424	Sh 151	54	Vir	10.8 − 18 34	7.0–7.5	34	5.31	1952	
425	STT 263			14.5 + 50 50	7.7–8.5	136	1.92	1965	
426	STF 1734			18.2 + 3 12	7.2–7.9	182	1.14	1963	
*427	STF 1744	ζ	UMa	21.9 + 55 11	2.1–4.2	150	14.54		N
428	STT 266			26.0 + 15 58	7.3–7.8	350	1.97	1966	
429	STF 1755			30.1 + 37 05	7.0–7.9	131	4.31		
430	STF 1757			31.7 − 0 04	7.8–8.9	114	2.19	1980 Orb.	
431	BU 932			32.0 − 12 57	6.1–6.6	46	0.35	1980 Orb.	
432	STF 1763	81	Vir	35.0 − 7 37	7.5–7.5	40	2.60	1958	
*433	STF 1768	25	CVn	35.2 + 36 33	5.7–7.6	101	1.72	1980 Orb.	
434	STF 1770			35.7 + 50 58	6.4–7.9	121	1.80	1967	
435	BU 612			37.1 + 11 00	6.0–6.0	188	0.23	1980 Orb.	N
436	STF 1772	1	Boo	38.3 + 20 12	6.2–9.1	137	4.57	1963	
437	STF 1777	84	Vir	40.6 + 3 47	5.8–8.2	227	3.01	1974	
438	STF 1781			43.6 + 5 22	7.8–8.2	96	0.32	1980 Orb.	N
439	STF 1785			46.8 + 27 14	7.2–7.5	159	3.41	1980 Orb.	
440	STF 1788			52.4 − 7 49	6.7–7.9	92	3.35	1971	N

230 Catalogue

1	2	3	4	5	6	7	8	9
441	STF 1793		$13^h56^m.8 + 26°04'$	7.0–8.0	243°	4".76	1956	
442	STT 276		14 06.1 + 36 59	7.5–8.3	204	0.60		
*443	STF 1821	κ^2 Boo	11.7 + 52 01	5.1–7.2	237	13.24		N
444	STF 1816		11.7 + 29 20	7.0–7.1	86	0.92	1966	N
445	KUI 66	15 Boo	12.4 + 10 20	5.5–8.1	119	1.11	1971	
446	STF 1819		12.8 + 3 22	7.9–8.0	239	0.82	1980 Orb.	
447	STF 1825		14.2 + 20 21	6.8–8.5	162	4.36	1962	
*448	STF 26	ι Boo	14.4 + 51 36	4.9–7.5	33	38.4		
449	STF 1834		18.5 + 48 44	7.1–7.2	104	1.24	1980 Orb.	
450	STF 1833		20.0 – 7 23	7.0–7.0	172	5.63		
*451	STF 1835		20.9 + 8 40	5.5–6.8	192	6.23		N
452	STF 1838		21.6 + 11 28	7.2–7.3	333	9.43		
453	STF 1837		22.0 – 11 27	7.1–8.7	283	1.22	1972	
454	STF 1846	ϕ Vir	25.6 – 2 00	5.2–9.7	111	5.00	1958	N
455	STF 1850		26.4 + 28 31	6.1–6.7	262	25.54		
456	STF 1863		36.4 + 51 48	7.1–7.4	68	0.64	1970	N
*457	STF 1864	π Boo	38.4 + 16 38	4.9–6.0	108	5.60		N
458	STF 1867		38.6 + 31 30	7.7–8.2	5	0.84	1974	
*459	STF 1865	ζ Boo	38.8 + 13 57	3.5–3.9	306	1.11	1980 Orb.	
460	STF 1871		39.8 + 51 37	7.0–7.0	304	1.81	1970	

231 Catalogue

1	2	3		4	5	6	7	8	9
*461	STF 1877	ε	Boo	$14^h42^m.8 + 27°17'$	3.0–6.3	338°	2″.79	1963	N
462	STF 1879			43.8 + 9 52	7.8–8.8	90	1.49	1980 Orb.	
463	BU 346			45.7 – 17 08	7.2–8.0	267	2.00	1952	
464	STF 1884			46.2 + 24 34	6.2–7.8	55	1.91	1963	
465	STF 1883			46.4 + 6 10	7.0–7.0	297	0.38	1980 Orb.	N
*466	BU 106	μ	Lib	46.6 – 13 57	5.4–6.3	358	1.83	1965	
467	A 1110			47.2 + 8 11	7.5–7.7	250	0.63	1970	
*468	STF 1890	39	Boo	48.0 + 48 55	5.8–6.5	45	2.96	1956	
*469	STF 1888	ξ	Boo	49.1 + 19 19	4.7–6.6	332	7.18	1980 Orb.	
470	STT 287			49.6 + 45 08	7.5–7.6	346	1.10	1980 Orb.	
471	STT 288			51.0 + 15 54	6.4–7.1	171	1.26	1980 Orb.	
472	BU 348	2	Ser	59.2 + 0 03	5.1–7.4	111	0.57	1962	
473	STF 1904			15 01.6 + 5 41	7.0–7.0	347	10.01		N
*474	STF 1909	44	Boo	02.2 + 47 51	5.2–6.1	29	0.90	1980 Orb.	
475	STF 1910			05.2 + 9 25	7.0–7.0	211	4.38	1971	
476	COU 189			09.8 + 19 10	6.0–7.9	141	0.50	1973	N
477	STF 1919			10.6 + 19 28	6.1–7.0	9	23.92		
478	STF 3091			13.4 – 4 43	7.7–7.7	229	0.45	1980 Orb.	
*479	STF 1932			16.2 + 27 01	5.6–6.1	250	1.40	1980 Orb.	
480	STF 1931			16.3 + 10 37	6.2–7.6	169	13.26		

232 Catalogue

1	2	3		4	5	6	7	8	9
481	BU 32	6	Ser	$15^h18^m.5 + 0°54'$	4.7–9.3	19°	3″.08	1964	
*482	STF 1937	η	CrB	21.1 + 30 28	5.2–5.7	321	0.39	1980 Orb.	N
*483	STF 1938	μ	Boo BC	22.6 + 37 31	6.7–7.3	15	2.17	1980 Orb.	N
484	STT 296			24.7 + 44 11	7.0–8.6	289	1.87	1963	
485	STF 1944			25.2 + 6 16	7.5–8.1	314	0.95	1970	
486	STF 1950			27.8 + 25 41	6.7–8.2	91	3.22	1966	
487	COU 610	θ	CrB	30.9 + 31 32	4.1–7	206	0.59	1975	N
*488	STF 1972	π^1	UMi	32.1 + 80 37	6.1–7.0	80	31.3		
*489	STF 1954	δ	Ser	32.4 + 10 42	3.0–4.0	178	3.95	1960	N
490	STT 298			34.3 + 39 58	7.0–7.3	211	0.76	1980 Orb.	N
491	STF 1963			35.9 + 30 16	7.3–7.7	296	5.02	1955	
492	STF 1962			36.0 – 8 38	6.3–6.4	189	11.82	1958	
*493	STF 1965	ζ	CrB	37.5 + 36 48	4.1–5.0	304	6.25	1961	
494	STF 1967	γ	CrB	40.6 + 26 27	4.0–7.0	126	0.39	1980 Orb.	N
495	BU 619			40.8 + 13 50	6.5–7.0	357	0.64	1960	
496	STF 1989	π^2	UMi	42.3 + 80 08	7.1–8.1	27	0.62	1980 Orb.	
497	STF 1984			49.7 + 53 03	6.2–8.5	275	6.50		
498	STF 1985			53.3 – 2 03	7.0–8.1	345	5.95	1955	
499	STF 1988			54.4 + 12 37	7.5–8.2	256	2.25	1963	
500	STF 2034			54.7 + 83 46	7.5–8.0	112	1.27	1974	

233 Catalogue

1	2	3		4	5	6	7	8	9
501	STT 303			15h58m.6 + 13°25'	7.4–7.9	164°	1".25	1962	N
*502	STF 1998	ξ	Sco	16 01.6 − 11 14	4.9–5.2	20	1.20	1980 Orb.	N
*503	H III 7	β	Sco	02.5 − 19 40	2.0–6.0	22	13.52		N
*504	STF 2010	κ	Her	05.8 + 17 11	5.0–6.0	12	28.2		
*505	BU 120	ν¹	Sco	09.1 − 19 21	4.2–6.7	2	1.24	1970	N
506	STF 2021	49	Ser	11.0 + 13 40	6.7–6.9	348	4.26	1971	
*507	STF 2032	σ	CrB	12.8 + 33 59	5.0–6.1	233	6.66	1980 Orb.	
508	STT 309			17.6 + 41 47	7.5–7.7	275	0.39		
509	Sh 228	5	Oph	22.6 − 23 20	5.0–6.0	352	3.33		
510	STF 2054			23.1 + 61 48	5.7–6.9	355	1.14	1968	
511	STT 312	η	Dra	23.3 + 61 38	2.1–8.1	143	5.30	1962	
512	STF 2049			25.8 + 26 06	6.5–7.5	200	1.21	1967	
*513	Antares	α	Sco	26.5 − 26 20	1.0–6.5	275	2.91	1974	N
514	STF 2052			26.7 + 18 31	7.5–7.5	134	1.38	1980 Orb.	
*515	STF 2055	λ	Oph	28.4 + 2 06	4.0–6.1	13	1.33	1980 Orb.	
516	STT 313			30.9 + 40 13	7.2–7.8	138	0.92		
*517	STF 2078	17	Dra	35.0 + 53 01	5.0–6.0	106	3.27	1966	
*518	STF 2084	ζ	Her	39.4 + 31 41	3.0–6.5	142	1.26	1980 Orb.	N
519	STF 2094			42.1 + 23 36	7.3–7.6	74	1.28	1966	
520	D 15			42.4 + 43 34	7.7–7.7	142	1.17	1980 Orb.	

234 Catalogue

1	2	3		4	5	6	7	8	9
521	MlR 198			16h43m.0 + 73°59'	7.3–7.6	214°	0".26	1971	N
522	STF 2106			48.7 + 9 30	6.7–8.4	180	0.55	1980 Orb.	
523	STF 2107			49.8 + 28 45	6.5–8.0	88	1.37	1980 Orb.	
*524	STF 2118	20	Dra	56.2 + 65 07	6.4–6.9	68	1.27	1980 Orb.	
525	STF 2114			59.6 + 8 31	6.2–7.4	182	1.34	1970	
*526	STF 2130	μ	Dra	17 04.3 + 54 32	5.0–5.1	42	1.90	1980 Orb.	N
*527	BU 1118	η	Oph	07.5 – 15 40	3.4–3.9	269	0.38	1980 Orb.	
528	STF 2135			10.0 + 21 17	7.1–8.4	187	7.85	1960	
*529	Sh 243	36	Oph	12.3 – 26 30	6.0–6.0	154	4.63	1980 Orb.	N
*530	STF 2140	α	Her	12.4 + 14 27	3.0–6.1	107	4.87	1971	N
*531	STF 3127	δ	Her	13.0 + 24 54	3.0–8.1	250	8.50	1969	N
532	A 2984	41	Oph	14.0 – 0 23	4.6–7.6	349	1.05	1974	
533	H III 25	39	Oph	15.0 – 24 14	6.0–7.0	355	10.8		
534	BU 126			17.0 – 17 42	6.4–7.5	261	2.01	1967	
*535	STF 2161	ρ	Her	22.0 + 37 11	4.0–5.1	316	3.96	1967	
536	STF 2168			24.9 + 35 48	7.5–8.2	199	2.39	1967	
537	STF 2180			27.8 + 50 55	7.0–7.2	261	3.07	1960	
538	STF 2173			27.8 – 1 01	5.8–6.1	349	0.49	1980 Orb.	N
539	STF 2186			33.3 + 1 02	7.5–7.5	79	2.98		
540	STF 2199			37.7 + 55 47	7.2–7.8	67	1.82	1968	

235 Catalogue

1	2	3	4	5	6	7	8	9
541	STF 2203		17ʰ39.7 + 41°41'	7.5–7.8	303°	0."76	1967	
542	STF 2218		40.0 + 63 42	6.5–7.7	327	1.80	1970	
543	HI 41		40.8 + 72 57	7.5–7.9	337	1.30	1968	
*544	STF 2202	61 Oph	42.0 + 2 36	6.2–6.5	94	20.41		
*545	STF 2241	ψ Dra	42.8 + 72 11	4.0–5.2	15	30.34		
546	STF 2213		43.0 + 31 09	7.5–8.0	329	4.51		
547	STF 2204		43.5 – 13 18	7.0–7.2	24	14.46		
548	STF 2215		44.9 + 17 43	5.9–7.9	275	0.61	1964	
549	STT 338		49.7 + 15 20	6.6–6.9	357	0.84	1967	
550	BU 130	90 Her	51.7 + 40 01	5.9–9.2	116	1.59	1966	
551	STF 2245		54.2 + 18 20	7.0–7.0	293	2.64		
552	STF 2271		59.2 + 52 51	7.3–8.3	266	2.89	1966	N
*553	STF 2264	95 Her	59.4 + 21 36	4.9–4.9	258	6.29	1967	N
*554	STF 2262	τ Oph	18 00.4 – 8 11	5.0–5.7	276	1.85	1980 Orb.	
*555	STF 2272	70 Oph	02.9 + 2 32	4.1–6.1	324	2.20	1980 Orb.	N
556	STF 2276		03.4 + 12 00	6.0–7.0	257	6.92		
557	STF 2282		04.9 + 40 21	7.2–8.2	85	2.46	1961	
558	AC 15	99 Her	05.1 + 30 33	6.0–9.0	4	1.46	1980 Orb.	
*559	STF 2280	100 Her	05.8 + 26 05	5.9–5.9	182	14.1		
560	STF 2289		07.9 + 16 28	6.0–7.1	224	1.17	1963	

Catalogue

1	2	3	4	5	6	7	8	9
561	BU 132		18ʰ08ᵐ3 − 19°52'	6.8–7.2	198°	1".44	1971	
562	STF 2294		12.0 + 0 10	7.4–7.7	93	1.02	1980 Orb.	
563	AC 11		22.4 − 1 36	7.0–7.2	356	0.78	1980 Orb.	
*564	STF 2323	39 Dra	23.2 + 58 46	4.7–7.7	352	3.72	1962	
*565	STF 2316	59 Ser	24.6 + 0 10	5.5–7.8	318	3.66	1962	
566	STF 2319		25.6 + 19 16	7.2–7.6	190	5.42	1975	
567	STT 354		29.6 + 6 45	7.2–8.0	195	0.73	1975	
568	A 1377		32.8 + 52 19	6.0–6.0	98	0.26	1980 Orb.	
569	STT 359		33.4 + 23 34	6.6–6.9	9	0.60	1980 Orb.	
570	STT 358		33.6 + 16 56	6.8–7.2	161	1.58	1980 Orb.	
571	STT 357		33.6 + 11 41	7.5–7.6	97	0.34	1980 Orb.	
572	STF 2368		37.7 + 52 18	7.2–7.4	144	1.80	1973	
573	STF 2369		41.4 + 2 34	7.5–8.0	76	0.53	1974	N
*574	STF 2382	ϵ^1 Lyr	42.7 + 39 37	4.6–6.3	358	2.54	1973	N
*575	STF 2383	ϵ^2 Lyr	id	4.9–5.2	92	2.20	1973	
576	STF 2375		43.0 + 5 27	6.2–6.6	118	2.51	1971	N
*577	STF 2379	5 Aql	43.9 − 1 01	5.6–7.4	121	13.0		
578	STF 2404		48.4 + 10 55	5.8–7.0	183	3.49	1966	
*579	STF 2420	o Dra	50.4 + 59 19	4.6–7.6	330	34.0		
580	STF 2415		52.4 + 20 33	6.6–8.5	291	1.98	1969	

237 Catalogue

1	2	3		4	5	6	7	8	9
*581	STF 2417	θ	Ser	18ʰ53.ᵐ7 + 4°08'	4.0–4.2	103°	22"24	1958	
582	STF 2422			55.1 + 26 02	7.6–7.7	80	0.64	1972	
583	STF 2438			56.6 + 58 09	7.0–7.6	2	0.89	1980 Orb.	N
584	Hd 150	ζ	Sgt	59.4 – 29 57	3.4–3.6	199	0.22	1980 Orb.	
585	STF 2449			19 04.0 + 7 05	7.1–7.8	290	8.16		N
586	STF 2455			04.8 + 22 06	7.2–8.3	36	3.02	1971	
587	STF 2486			10.8 + 49 45	6.0–6.5	212	8.23	1961	
*588	STF 2487	η	Lyr	12.0 + 39 03	4.0–8.1	83	28.18		
589	STT 368			13.8 + 16 04	7.3–8.5	218	1.12	1968	
590	STT 371			13.9 + 27 22	6.8–6.9	159	0.88	1963	
591	BU 248	2	Vul	15.6 + 22 56	5.7–9.5	127	1.78	1965	
592	STF 2492	23	Aql	16.0 + 1 00	5.5–9.5	4	3.08	1958	
593	STF 2509			16.4 + 63 07	7.0–8.1	331	1.58	1969	
594	STF 2525			24.5 + 27 13	7.4–7.6	291	1.68	1980 Orb.	
595	STF 2523			24.6 + 21 03	7.3–7.4	148	6.43	1953	
*596	STF 43	β	Cyg	28.7 + 27 51	3.0–5.3	54	34.34	1956	N
597	A 713			29.8 + 47 22	6.9–7.4	264	0.38	1972	
598	STF 2571			31.8 + 78 09	7.3–8.0	19	11.38	1962	
599	STT 378			34.9 + 40 54	7.2–9.0	287	1.42	1962	
600	STF 2545			36.0 – 10 16	6.2–8.1	322	3.74	1969	

238 Catalogue

1	2	3		4	5	6	7	8	9
601	STF 2556			19ʰ37ᵐ.3 + 22°08'	7.3–7.8	36°	0".39	1980 Orb.	
602	STT 382			39.8 + 27 16	7.1–7.6	332	0.36	1970	
603	STT 380			40.2 + 11 42	6.0–7.2	75	0.47	1962	
*604	STF 46	16	Cyg	40.6 + 50 24	5.1–5.3	134	38.5		
605	STT 384			42.0 + 38 12	7.0–7.3	195	0.94	1967	
*606	STF 2579	δ	Cyg	43.4 + 45 00	3.0–7.9	233	2.29	1980 Orb.	
607	STF 2576			43.6 + 33 30	7.8–7.8	356	2.03	1980 Orb.	
608	STF 2578			43.8 + 35 58	6.6–7.4	125	14.78	1967	
609	STF 2580	17	Cyg	44.5 + 33 37	5.1–8.1	70	25.91	1960	
*610	STF 2583	π	Aql	46.4 + 11 41	6.0–6.8	108	1.51	1967	
611	STT 387			46.8 + 35 11	7.2–8.2	166	0.60	1980 Orb.	
*612	STF 2603	ε	Dra	48.3 + 70 08	4.0–7.6	14	3.10	1963	
613	STT 388			50.2 + 25 44	7.6–7.6	139	3.84	1954	
614	STF 2596			51.7 + 15 10	7.2–8.6	307	1.93	1970	
*615	STF 2605	ψ	Cyg	54.4 + 52 18	5.0–7.5	177	3.10	1962	
616	AC 12			55.8 – 2 22	7.0–8.0	307	1.38	1970	
617	STF 2606			56.6 + 33 08	7.5–8.2	141	0.82	1972	
618	STF 2609			56.8 + 37 58	7.0–8.1	23	1.94	1966	
619	STF 2613			59.0 + 10 36	7.0–7.2	352	3.92	1967	
*620	STT 395	16	Vul	59.9 + 24 48	5.8–6.2	119	0.79	1969	

239 Catalogue

1	2	3	4	5	6	7	8	9
621	STF 2624		$20^h01^m.6 + 35°53'$	7.2–7.8	175°	1".89	1969	
622	STF 2621		02.2 + 9 06	7.7–7.9	224	5.67		
623	STF 2628		05.4 + 9 15	6.1–8.2	342	3.56	1958	
624	STF 2644		10.0 + 0 43	7.1–7.4	207	2.82	1967	
*625	STF 2675	κ Cep	10.7 + 77 34	4.0–8.0	122	7.43		
626	STF 2655		11.9 + 22 04	7.5–7.5	182	6.23	1956	
627	STT 403		12.6 + 41 57	7.0–7.2	172	0.89	1959	
628	BU 984		15.5 + 26 13	7.9–8.2	238	0.68	1967	
629	STF 2666		16.4 + 40 35	6.5–8.7	246	2.63	1965	
630	STF 2671		17.2 + 55 14	6.0–7.4	338	3.30	1959	
631	STT 406		18.2 + 45 12	7.1–8.0	117	0.58	1980 Orb.	
632	A 730		24.0 + 59 26	6.8–7.0	322	0.23	1980 Orb.	
*633	BU 60	π Cap	24.5 − 18 22	5.1–8.7	149	3.28	1966	N
634	STT 407		28.8 + 11 05	7.0–7.5	255	16.75	1964	N
635	STF 2695		29.8 + 25 38	6.2–8.0	90	0.40	1975	N
*636	BU 151	β Del	35.2 + 14 25	4.0–5.0	5	0.59	1980 Orb.	
637	STF 2705		35.8 + 33 11	7.1–8.1	262	3.16	1961	
638	STF 2717		36.8 + 60 35	7.2–9.7	260	1.99	1965	
639	STT 410		37.7 + 40 24	6.4–6.7	9	0.85	1967	
640	STF 2716	49 Cyg	39.0 + 32 08	6.0–8.1	47	2.64	1969	

240 Catalogue

1	2	3		4	5	6	7	8	9
641	STF 2718			20^{h}40^m.2 + 12°33'	7.4-7.6	86°	8".34	1969	
642	BU 152			41.0 + 57 12	7.2-8.0	88	0.98	1969	N
643	STF 2723			42.5 + 12 08	6.4-8.2	116	1.18	1960	N
644	STF 2726	52	Cyg	43.6 + 30 32	4.0-9.2	65	6.39	1967	
645	STF 2725			43.9 + 15 43	7.3-8.0	8	5.77	1967	N
*646	STF 2727	γ	Del	44.4 + 15 57	4.0-5.0	268	10.02	1956	N
647	STT 413	λ	Cyg	45.5 + 36 18	5.0-6.3	25	0.84	1967	
648	S 763			45.6 − 18 23	6.5-7.2	293	15.80		
649	STF 2730			48.6 + 6 12	7.8-7.9	335	3.31	1967	
650	STF 2729	4	Aqr	48.8 − 5 49	6.3-7.6	11	1.05	1975	
651	Ho 144			50.1 + 19 56	7.0-7.0	347	0.29	1971	
652	STT 416			50.2 + 43 34	7.8-8.1	125	8.41	1960	
653	STT 417			51.0 + 28 57	7.5-8.1	29	0.85	1967	
654	STT 418			52.8 + 32 31	7.3-7.4	286	1.14	1967	
655	STF 2735			53.2 + 4 20	6.2-7.7	286	1.99	1965	
656	STF 3133			56.5 + 61 10	7.4-8.9	101	3.26	1972	
657	STF 2737	ε	Equ	56.6 + 4 06	5.7-6.2	285	1.07	1980 Orb.	
658	STF 2741			56.9 + 50 16	6.0-7.3	27	1.91	1957	
659	STF 2742	2	Equ	59.8 + 6 59	7.1-7.1	219	2.62	1964	
*660	STF 2744			21 00.5 + 1 20	6.5-7.0	125	1.27	1980 Orb.	N

Catalogue

1	2	3		4	5	6	7	8	9
*661	STF 2751			$21^h00^m.7 + 56°28'$	6.0-7.0	353°	1".61	1962	
662	STF 2745	12	Aqr	01.4 − 6 01	5.6-6.7	193	2.68	1971	
*663	STF 2758	61	Cyg	04.4 + 38 28	5.6-6.3	146	29.04	1980 Orb.	N
664	STF 2760			04.7 + 33 56	7.3-8.1	15	1.27	1975	
665	STF 2762			06.5 + 30 00	6.0-8.0	308	3.30	1962	
666	STF 2767			08.2 + 19 45	7.8-8.2	30	2.39	1966	
667	STF 2769			08.3 + 22 15	6.5-7.5	300	17.89		
668	STF 2765			08.6 + 9 21	7.8-8.0	81	2.68	1965	
669	STT 431			09.6 + 41 02	7.6-8.0	123	2.85	1964	
670	STF 2780			10.5 + 59 47	6.2-7.2	218	1.11	1956	
671	STT 535	δ	Equ	12.0 + 9 48	5.3-5.4		dist. var.		N
672	STT 432			12.4 + 40 56	6.8-7.2	120	1.32	1956	
673	AGC 13	τ	Cyg	12.8 + 37 49	3.8-8.0	144	0.82	1980 Orb.	N
674	STF 2786			17.2 + 9 19	7.0-8.1	189	2.75	1965	
675	STT 437			18.7 + 32 14	6.5-7.2	27	2.14	1962	
676	STT 435			18.9 + 2 40	7.5-8.0	229	0.66	1971	
677	STF 2801			20.2 + 80 08	7.3-8.0	272	1.88	1970	
678	Ho 157			20.9 + 31 49	7.7-7.7	22	4.26		
679	COU 430			22.8 + 18 15	7.5-8.4	232	0.58	1975	
680	STF 2797			24.3 + 13 28	6.7-8.2	216	3.19	1965	

242 Catalogue

1	2	3	4	5	6	7	8	9
681	STF 2799		21h26m.4 + 10°52'	7.0–7.0	273°	1".68	1967	
682	STF 2806	β Cep	28 .0 + 70 20	3.3–8.0	249	13.42	1962	
683	STF 2804		30 .6 + 20 29	7.3–8.0	349	3.15	1967	
684	Hu 371		33 .2 + 24 14	7.0–7.5	295	0.30	1980 Orb.	
*685	STF 2822	μ Cyg	41 .9 + 28 31	4.7–6.1	298	1.81	1980 Orb.	
686	BU 989	κ Peg	42 .4 + 25 25	4.8–5.3	293	0.20	1980 Orb.	N
*687	STF 2840		50 .3 + 55 33	6.0–7.0	196	18.33	1958	
688	STF 2843		50 .4 + 65 31	7.0–7.2	145	1.56	1975	
689	STT 456		53 .7 + 52 17	7.8–8.0	35	1.48	1960	
690	STT 457		54 .2 + 65 04	6.3–8.5	245	1.14	1967	
691	STT 458		54 .9 + 59 32	7.1–8.6	350	0.84	1965	
692	STF 2847		55 .5 – 3 43	7.6–8.0	304	0.88	1971	
693	BU 276	η PsA	58 .0 – 28 42	5.0–6.0	115	1.98		
694	STF 2873		22 00 .4 + 82 38	6.2–7.0	72	13.85		
695	STF 2854		02 .0 + 13 24	7.7–8.0	84	1.94	1974	
*696	STF 2863	ξ Cep	02 .2 + 64 23	4.6–6.5	278	7.60	1962	N
697	STF 2862		04 .5 + 0 19	7.6–8.0	98	2.47	1970	
698	COU 136		07 .7 + 22 53	7.6–7.6	52	0.38		
699	Sh 339	41 Aqr	11 .5 – 21 19	5.7–7.7	116	4.96	1974	N
*700	STF 2893		12 .0 + 73 04	5.5–7.6	348	29.1		N

243 Catalogue

1	2	3	4	5	6	7	8	9
701	STF 2878		$22^h12^m.0 + 7°44'$	6.5-8.0	121°	1".42	1962	
702	STF 2881		12.3 + 29 20	7.7-8.2	85	1.37	1967	
703	Ho 180		13.7 + 43 39	7.2-7.2	236	0.68	1961	
704	STF 2903		20.2 + 66 27	7.0-8.0	96	4.19	1970	
705	BU 172		21.5 − 5 06	6.7-6.7	277	0.27	1980 Orb.	N
706	Sh 345	53 Aqr	23.8 − 17 00	6.3-6.6	327	3.92	1966	
*707	STF 2909	ζ Aqr	26.2 − 0 17	4.4-4.6	229	1.63	1980 Orb.	N
*708	STF 1912	37 Peg	27.4 + 4 11	5.8-7.2	118	1.04	1980 Orb.	
709	Hu 981		28.8 + 61 22	7.5-7.7	226	0.35	1974	
710	STF 2924		31.6 + 69 39	6.8-7.3	84	0.56	1980 Orb.	
711	A 1468		32.2 + 53 50	7.7-7.7	79	0.27	1970	
*712	STF 2922	8 Lac	33.6 + 39 23	6.0-6.5	185	22.44	1958	
*713	Ho 296		38.4 + 14 17	5.5-5.5	25	0.27	1980 Orb.	N
714	STF 2935		40.4 − 8 34	7.0-8.0	310	2.44	1962	
715	STT 529		43.4 + 67 52	7.5-8.8	202	3.68	1966	N
716	STF 2944		45.3 − 4 29	7.0-7.5	278	2.47	1967	N
717	STF 2947		47.3 + 68 18	7.2-7.2	60	4.22	1962	N
718	STF 2948		47.8 + 66 17	7.0-8.7	3	2.50	1965	
719	Ho 482		49.0 + 26 08	6.8-6.8	39	0.34	1980 Orb.	
720	STF 2950		49.4 + 61 25	5.7-7.0	292	1.64	1970	

244 Catalogue

1	2	3	4	5	6	7	8	9
721	STF 2963		$22^h53^m.0 + 76°04'$	7.8–8.5	357°	1".95	1965	N
722	COU 240		54.0 + 22 41	7.4–7.8	289	0.64	1973	
723	STF 2971		55.7 + 78 13	7.3–8.5	4	5.50	1962	
724	STT 483	52 Peg	56.7 + 11 28	6.2–7.7	300	0.68	1980 Orb.	
725	BU 180		23 05.0 + 60 33	7.5–8.0	156	0.64	1957	
*726	STT 489	π Cep	06.3 + 75 07	5.2–7.5	334	0.96	1980 Orb.	
727	BU 385		07.9 + 32 13	7.1–7.9	97	0.62	1974	
728	STF 2988		09.4 − 12 13	7.2–7.2	101	3.51	1961	
*729	STF 3001	o Cep	16.4 + 67 50	5.2–7.8	217	2.90	1980 Orb.	
*730	STF 2998	94 Aqr	16.4 − 13 44	5.2–7.2	350	12.97	1958	
731	STF 3008		21.2 − 8 44	7.0–8.0	170	4.02	1968	
732	STF 3017		25.7 + 73 49	7.1–8.2	25	1.56	1970	
*733	BU 720	72 Peg	31.5 + 31 03	6.0–6.0	80	0.51	1980 Orb.	
734	STT 500		35.1 + 44 09	6.1–7.0	353	0.48	1971	
735	STT 503		39.5 + 20 01	7.2–7.8	132	1.16	1973	
736	AGC 14	78 Peg	41.4 + 29 06	5.0–8.1	238	0.97	1970	
*737	Sh 356	107 Aqr	43.4 − 18 57	5.3–6.5	137	6.61	1962	
738	STF 3037		43.6 + 60 12	7.0–8.5	211	2.70	1965	
739	STT 507		46.2 + 64 36	6.8–7.5	305	0.71	1974	
740	STT 508	6 Cas	46.4 + 61 57	5.7–8.2	196	1.55	1964	
741	STT 510		49.0 + 41 48	7.5–7.8	131	0.49	1971	
742	STF 3042		49.3 + 37 37	7.0–7.0	87	5.42	1953	
*743	STF 3049	σ Cas	56.4 + 55 29	5.4–7.5	326	3.09	1954	
*744	STF 3050		56.9 + 33 27	6.5–6.5	309	1.55	1980 Orb.	N

Notes to the Catalogue

No. 11	The separation is increasing: 1861 276° 0."57.
No. 12	Large orbital motion: 1889 54° 0."39, 1920 84° 0."44.
No. 19	Orange and sapphire blue (Flammarion).
No. 25	Visible in the smallest instruments. Yellow and lilac (Flammarion).
No. 31	Requires a 25-cm objective. The separation is decreasing: 1876 92° 0."75.
No. 34	The components are separating, but the object is resolvable only with large objectives: 1917 10° 0."18.
No. 39	Good test for an aperture of 40 cm. Large orbital motion: 1887 196° 0."25, 1923 220° 0.29. Since 1940 the quadrant has often been reversed.
No. 45	The separation is increasing: 1875 74° 0."48.
No. 48	Resolvable in the smallest refractors. The companion is double, itself having a magnitude-11 companion at 1."18 around 233°.
No. 51	The companion is a very close double at 0."13.
No. 56	Good seeing and an excellent objective are required. The components are approaching each other: 1875 289° 1."34.
No. 59	Closing in: 1831 88° 4."01.
No. 67	The Pole Star. The companion can be seen in a 6-cm refractor, but to do so requires some practice.
No. 71	Very easy. Closing in: 1830 0° 8."63.
No. 77	Orange and emerald (Flammarion). One of the most beautiful systems.
No. 78	Together with the preceding one, this forms a very difficult triple system for average instruments. Flammarion noted that the components are green and blue. Observation requires excellent seeing.
No. 89	A companion of magnitude 8.1 at 7."2. Golden yellow, lilac, and purple (Flammarion).
No. 90	Discovered in 1937. Typical test for a 20-cm mirror.
No. 108	A companion of magnitude 10.5 at 25.".
No. 112	The primary is a very close double: 6^m9"–9^m2, 0."15.
No. 113	Closing in: 1832 14° 3."59.
No. 115	Becoming less and less difficult: 1830 189° 0."55.
No. 118	The separation is decreasing: 1875 87° 2."72.
No. 121	Becoming easier: 1831 146° 1."34.
No. 129	Discovered in 1968. Try observing it with apertures of 30 cm or more, in good seeing.
No. 132	The primary was discovered to be double in 1968: 7^m4"–8^m1, 20° 0."27.
No. 140	Closing in: 1830 105° 2."87.
No. 142	The companion is double: 7^m2"–9^m8, 255° 0.66 1962.
No. 150	Discovered in 1937.
No. 159	This double star is in the middle of the cluster H VII 47.

No. 176	Discovered in 1938.
No. 187	Discovered in 1934. Direct orbital motion with decreasing separation: 1935 1° 1"13.
No. 188	The primary is a double resolvable only with the largest instruments: 1980 $5^m.4-7^m.4$ 144° 0".23 Orbit.
No. 189	1829 99° 1".58.
No. 194	Separation increasing: 1843 227° 2".66.
No. 199	The companion is double: 1980 $8^m.7-9^m.2$ 77° 0".35 Orbit.
No. 204	The separation is smaller than the orbit indicates: 0".45±.
No. 206	Rigel. The companion can be seen with a small refractor. The companion is itself double, with a separation of 0".1.
No. 209	1835 273° 1".42.
No. 213	The companion is double, but practically unobservable for a number of years: 1980 322° 0".10 $8^m.0-8^m.1$ Orbit.
No. 217	Becoming very difficult. It now requires a 30-cm instrument. 1854 80° 0".82.
No. 220	There is another double in the field, 30" away: $8^m.6-9^m.7$, 305° 0".65.
No. 221	Discovered in 1972.
No. 223	This is the Trapezium of Orion, in the nebula. These are young stars, born less than a million years ago, in the bosom of the nebula.
No. 228	The separation is increasing. The system is becoming a good test for a 95-mm objective. 1829 23° 0".67.
No. 229	The primary is a very close double, 0".1–0".2.
No. 238	Opening out: 1871 6° 2".15.
No. 243	Discovered in 1934.
No. 244	Requires a good objective. Try it with one of 162 mm aperture. The primary is variable from $3^m.3$ to $4^m.2$ in 230 days.
No. 247	Direct motion: 1830 89° 0".81.
No. 251	The companion is a very easy double: $5^m.5-6^m.0$, 107° 2".89.
No. 258	Becoming difficult: 1830 249° 1".06.
No. 259	Discovered in 1972. Nevertheless, resolvable with a 15-cm aperture.
No. 262	Closing in: 1846 137° 30".40.
No. 263	Good test for a 75-mm instrument. Try it with one of 61 mm. At 8".6 there is a $7^m.4$ component.
No. 264	The companion of Sirius was the first known white dwarf. One can try to see it until 1984 with instruments of at least 20 cm aperture. After that it will be too close to the primary star, and invisible even in the most powerful instruments until the year 2000.
No. 269	According to Flammarion the companion varies in brightness and color. While the primary is golden yellow, the companion changes from green to blue, purple, and red. Retrograde motion: 1829 175° 5".73.
No. 270	Closing in: 1831 149° 3".67.
No. 275	Widening out: 1830 159° 2".94.

No. 277   The orbit by Karmel gives separations that are too large.
No. 282   Direct motion: 1829 197° 7″.14.
No. 283   Direct motion: 1831 115° 0″.48.
No. 289   Direct motion: 1831 292° 2″.35.
No. 291   One of the brightest pairs in the sky. A good test for small refractors. Rapid orbital motion: 1985 87° 2″.68 Orbit.
No. 295   In the field of Procyon. Typical pair for a 10-cm aperture: 1829 132° 1″.46.
No. 300   Discovered in 1937. Retrograde motion: 1937 99° 0″.18. This is the primary of BU 1062 6^m–$13^m\!.5$ at 4″.2 36°. Try the system with reflectors of 30–40 cm.
No. 301   Closing in. Becoming difficult for small refractors: 1831 17° 3″.33.
No. 303   Resolvable only with large apertures. The orbital motion is rapid, with large changes in the separation: 1985 113° 0″.15, 1990 292° 0″.60 Orbit.
No. 305   Closing in: 1831 267° 1″.59.
No. 306   Separation decreasing: 1831 205° 2″.37.
No. 311   Opening up: 1829 71° 1″.61.
Nos. 312 and 313   Together these form a famous triple system. The motion of the close pair is rapid: 1985 239° 0″.66 Orbit.
No. 315   The companion is a very close double in very rapid motion: 1980 276° 0″.16 Orbit.
No. 322   Opening up: 1892 184° 0″.39.
No. 324   Pale orange and clear blue (Flammarion). Look at it with a pair of binoculars.
No. 325   Yellow and blue. The primary is a very difficult double: $4^m\!.0$–$5^m\!.5$ 1980 155° 0″.25 1980 Orbit.
No. 334   The companion is double: $9^m\!.5$–$9^m\!.8$ 1980 201° 0″.92 Orbit.
No. 335   Discovered in 1936. This pair is always difficult, even at its greatest separation. It is in rapid motion: 1985 326° 0″.54 Orbit.
No. 337   For large apertures. Rapid orbital motion: 1985 275° 0″.22 Orbit.
No. 342   Closing in: 1830 245° 3″.48.
No. 345   Rapid orbital motion: 1985 212° 0″.48, 1990 262° 0″.20 Orbit.
No. 356   Opening out: 1909 35° 0″.80.
No. 361   Discovered in 1935. Try it with an aperture of 30 cm.
No. 363   Very close binary in rapid motion: 1985 137° 0″.20.
No. 369   One of the most beautiful pairs in the sky. Two yellow translucid diamonds (Flammarion). Resolvable with small naval telescopes.
No. 371   Binary in rapid motion. Large changes in the separation: 1985 231° 0″.39 Orbit. Not easy.
No. 373   Opening out: 1829 288° 0″.71.
No. 375   Good test for an aperture of 162 mm. Retrograde motion: 1846 347° 0″.68.
No. 382   The primary is yellow. The companion is variable, according to Flammarion its colors change from blue to yellow, to indigo and

	purple. Should be observed in good seeing; rarely visible with an aperture of less than 15 cm.
No. 385	White and pale rose (Flammarion). Look at the couple with a pair of binoculars.
No. 387	For objectives of at least 30 cm aperture. Binary in rapid motion: 1985 135° 0".37 Orbit.
No. 388	As difficult as the preceding one. Binary in rapid motion: 1985 256° 0".53 Orbit.
No. 393	Opening out: 1847 287° 0".74.
No. 405	Orange and lilac (Flammarion). To be looked at with a naval telescope.
No. 407	One of the brightest pairs in the sky: 1985 292° 3".49 Orbit.
No. 411	A star of magnitude 9 at 28".
No. 412	Cor Caroli. One of the prettiest doubles in the sky according to Flammarion; golden yellow and lilac. Look at it with a small telescope.
No. 418	Discovered in 1959. The separation is increasing: 1959 321° 0".99.
No. 422	Binary in rapid motion. The orbital plane contains the observer: 1985 192° 0".61 Orbit.
No. 423	Opening out and becoming fairly easy: 1843 359° 0".63.
No. 427	Mizar, the second star in the handle of the Big Dipper. Nearby, at 11'48", is the magnitude-5 star Alcor, visible to the naked eye. It is good to look at Mizar and Alcor in the same field, as one can with the smallest refractors, to compare the separations.
No. 435	For large apertures. Binary in rapid motion: 1985 213° 0".32 Orbit.
No. 438	1985 126° 0".41.
No. 440	Easy. Opening up: 1831 54° 2".36.
No. 444	Closing in again: 1831 80° 1".87.
No. 445	Discovered in 1939.
No. 451	The companion is a difficult double in rapid orbital motion: 1980 26° 0".28, 1985 50° 0".26 Orbit.
No. 454	Opening up: 1829 109° 3".73.
No. 456	Slow orbital motion: 1830 110° 0".65.
No. 458	Closing in: 1831 22° 1".63.
No. 461	"Pulcherrima." Bright yellow and marine blue (Flammarion). Try it with 75-mm refractor, or even with one of 61 mm. Very slow motion: 1829 321° 2".64.
No. 465	Rapid change of separation: 1985 291° 0".53 Orbit.
No. 474	Becoming accessible again to small instruments. The separation is increasing rapidly: 1985 39° 1".25 Orbit.
No. 476	Discovered in 1967. Requires very good seeing.
No. 482	The pair is difficult now, but the separation is increasing: 1985 10° 0".80 Orbit.
No. 483	The principal star of μ Bootis is of magnitude 4 at 108" from the pair BC.

No. 487 A very bright pair discovered in 1971. It can be seen with an aperture of 30 cm.
No. 489 The separation is increasing: 1833 197° 2".66.
No. 490 Becoming difficult: 1985 229° 0".47 Orbit.
No. 494 The companion is separating, but it is still an object for the most powerful instruments: 1985 121° 0".53 Orbit.
No. 501 The separation is increasing: 1846 111° 0".60.
No. 502 A $7^m\!.2$ companion at 54° and 7".7.
No. 503 The primary is a very difficult double even in the largest instruments: $2^m\!.0$–$9^m\!.7$, 105° 0".77.
No. 505 The component ν^2 at 40" is double itself: $7^m\!.0$–$8^m\!.0$, 54° 2".27. It is a very pretty sight with an objective of at least 10 cm. It resembles ϵ Lyrae.
No. 513 Antares. Orange and green. No appreciable motion since 1847. Observe it during twilight some summer evening.
No. 518 Really difficult. Requires an excellent objective and a calm atmosphere: 1985 110° 1".44 Orbit.
No. 521 Discovered in 1971.
No. 527 Good test for a 25-cm mirror: 1925 260 0".44 Orbit.
No. 529 Easy. Has a common proper motion with 30 Sco 14' away and magnitude 7.
No. 530 Orange and emerald (Flammarion). The primary is variable. For small refractors.
No. 531 Clear blue and violet (Flammarion).
No. 538 Rapid orbital motion: 1985 339° 0".83 Orbit.
No. 552 Opening out: 1831 262° 1".88.
No. 553 One cannot do better than to quote Flammarion: "Golden yellow and light azure, an extremely pretty pair, very bright, a charming little picture, variable colors" (*Les Étoiles*, p. 226).
No. 555 Very fine in a small refractor. Significant orbital motion: 1980 287° 2".12 Orbit.
No. 573 Becoming difficult because the components are approaching each other: 1830 98° 1".54.
No. 574 Those with very good eyesight can resolve ϵ^1 and ϵ^2 (at 207") with the naked eye. A 56-mm refractor is sufficient to show the sight.
No. 576 The two components are very close doubles at 0".1.
No. 584 The separation varies considerably, the motion is very rapid: 1985 72° 0".45 Orbit.
No. 586 Large motion, with the separation getting smaller: 1828 144° 4".93.
No. 596 Albireo, one of the most beautiful pairs in the sky. Has appeared fixed since 1832. Golden yellow and sapphire (Flammarion). Look at it with a pair of binoculars.
No. 634 The companion is a double becoming more and more difficult to see: 1846 212° 0".57, 1975 256° 0".15.

No. 635 Closing in after having passed through the extremity of the apparent major axis: 1831 76° 0.″80, 1883 78° 1.″16, 1923 79° 0.″91.
No. 636 Rapid orbital motion: 1985 45 0.″25 Orbit.
No. 642 Retrograde motion with increasing separation: 1876 111° 0.″45.
No. 643 The primary is a very close double.
No. 645 The separation is increasing: 1829 358° 4.″24.
No. 646 Orange and green. Separation decreasing: 1830 274° 11.″90.
No. 657 Good test for a 10-cm refractor. A 7.^{m}1 component at 10.″2.
No. 664 Large rectilinear motion, opening up again:

1829	223°	13.″66
1888	225°	7.″74
1924	230°	3.″88
1940	236°	2.″20
1960	291°	0.″75
1965	334°	0.″59

No. 671 Only resolvable with the most powerful objectives. This pair is a regular clock; its period is 5.7 years. Here is an ephemeris:

1979.7	16°	0.″26
1980.2	7°	0.″14
1980.7	248°	0.″05
1981.2	204°	0.″15
1981.7	183°	0.″09
1982.2	60°	0.″07
1982.7	36°	0.″19
1983.2	30°	0.″29
1983.7	27°	0.″34

No. 673 Becoming more and more difficult. Look at it in perfect seeing with a large instrument: 1985 102° 0.″54 Orbit.
No. 679 Discovered in 1969.
No. 686 Binary in rapid motion: for mirrors of at least 50 cm aperture: 1985 134° 0.″22 Orbit.
No. 696 Very easy. The separation is increasing: 1831 289° 5.″60.
No. 698 Discovered in 1966, the separation was 0.″24. Try it with a 25-cm mirror.
No. 699 Yellow topaz and sky blue (Flammarion).
No. 705 Orbital motion: 1985 259° 0.″23 Orbit.
No. 707 Good test for small refractors. Orbital motion: 1985 208° 1.″65 Orbit.
No. 713 Rapid orbital motion: 1985 112° 0.″13 Orbit.
No. 715 A 9.^{m}0 component at 20.″71.
No. 716 Closing in: 1832 247° 4.″12.
No. 717 Opening out: 1832 76° 2.″98.
No. 722 Discovered in 1967. Fairly easy. Try it with a 20-cm aperture.
No. 744 Good test for a 75-mm refractor. 1985 316° 1.″60 Orbit.

INDEX

Absorption, 81, 82
AC, 105
ADS, 158
Africano, J., 70
AGK1, 95, 96
AGK2, 96, 100
AGK3, 96, 107, 190
Airy, George B., 7
Airy disk, 3, 32, 33, 35, 37, 40, 56
Airy figure, 31
Aitken, Robert G., 6, 10–12, 19, 20, 95, 158
Algol, 3, 5
Allen, L. R., 70
Ann Arbor Observatory, 11
Anomalies
 mean, 113, 116, 138, 140, 142
 true, 111, 116, 119, 120, 139, 140, 142
Antares, 5, 156
Apostron, 110, 135
Arago, D. F., 6
Arcturus, 174
Areas, law of, 124, 129, 134
Arend, S., 19
Argelander, F., 13, 20, 96–98, 190
Artificial double stars, 89–90
Artificial satellites, 96, 178
Astrometric double stars, 4, 15, 64, 68–69
Astronomische Gesellschaft, 10, 93, 94–95
Atlas (binary), 71

Atmospheric distortion, xiv, 13, 40, 53–54, 66, 77–86
Auwers, 93
Auzout, 48

Baize, Paul, xiv, 6, 16–19, 56, 122, 123, 166–170, 202
Balmer lines, viii
Barney, Ida, 95
BD, 20, 96, 98–106
Belgrade Observatory, 13
Bessel, Friedrich, 94
Betelgeuse, 156
Bloemfontein Observatory, 6, 11, 14, 22
Blurring, 82–85
Bootis
 ϵ, 5
 ζ, 40
 τ, 5–6
Boss, B., 95
Bradley, James, 94
Brightness. *See* Luminosity
Brown, R. Hanbury, 70
Burnham, Sheldon W., 6, 9–12, 94

Cameras. *See* Photography
Campbell elements, 113–119
Cancri, ζ, 32, 40, 199
Capella, 156, 158, 176
Cannon, 157
Carte du Ciel, 14–15, 21, 97–106

Cassegrain arrangement, 32
Cassiopeiae, η, 21
Castor, 7, 146, 177, 199
Catalogue des étoiles doubles et multiples en mouvement relatif certain (Flammarion), 16
Catalogue of Nearby Stars (Gliese), 106
Catalogues, 8, 10–13, 16, 19–20, 92–93, 158. *See also* names and abbreviations of catalogues
  fundamental, 93
  general, 97–106
  intermediate, 94–96
  of red dwarfs, 19, 106
  special, 106–107
Catalogus novus stellarum duplicium (Struve), 8, 9
Celestial pole, 121
Centauri, α, 2, 4, 170, 175–176
Central card index, 20, 22
Centre d'Etudes et de Recherches Géodynamiques et Astronométriques, 70
Cephei, β, 69
Characteristic curves, 135–139
Charts, celestial, 10–11
Chromatic aberration, 31, 32, 86
  lateral, 53
Clark, Alvan, 141
Classification of stars, 4, 12–13, 155–156
Clocks, 85, 99
Close double stars, 34–40, 70–71, 190, 191
Comae, 39, 190
Comets, 42, 94
Comparators, 51, 53
Coordinates, 92–93, 97–105, 116, 158
Cordoba catalogue, 20
Coronae Borealis, θ, 32–33, 190
Couder, A., 28, 34
COU 14, 199

Danjon, André, ix, xiii, 19, 28, 34, 57, 59, 78, 82, 86, 116, 122, 133–145, 171
Davidson, 56
Davis, J., 70
Dawes, W. R., 16
Dearborn Observatory, 9, 107
Declination, 81, 98–100, 106
Dembowski, Ercole, 16
Dependences, method of, 155
Dieckvoss, W., 96
Differential correction, 145–148
Differential micrometric effect, 61–62
Diffraction, vii, 34
Diffraction image. *See* Airy figure
Diffusion, 81, 82
Dispersion, 81, 82
Diurnal motion, 47, 65, 66
DM, 97
Domes, 78–79, 85
Dommanger, Jean, 19
Doolittle, E., 11
Doppler-Fizeau effect, vii–viii, 4
Dorpat Observatory, 8, 9
Draconis, χ, 69
Dunham, D., 70
Duruy, V., 19, 56, 66
Dwarf stars, xv, 12–13, 192–194
  red, 12, 13, 15, 19, 21, 106, 141, 175, 177, 182, 184
  white, 15, 158, 166, 175
Dynamical elements, 113
Dynamical parallaxes. *See* Parallaxes

Eclipses, 3–5
Ecliptic, 93
Eddington, A., 166
Electronography, 66
"Empirical Data on Stellar Masses, Luminosities, and Radii," 158
Ephemeris, 115–116, 119, 121
Equator, 93

Index

Espin, T. E. H., 11, 12, 202
Evans, D., 70
Exit pupil, 30, 53, 86

Far-sighted observers, 29–30
FC, 93
Finsen, W. S., 19, 57
Fizeau-Michelson interferometer, 56–57, 59, 67
FK3, 93
FK4, 93
Flagstaff Observatory, 15, 71, 106, 155
Flammarion, Camille, xiii–xiv, 6, 16
Flamsteed, John, 94
Flemming, 157
Fourier transforms, 67–68, 72
Franz, Otto, 71
Fraunhofer, Johann von, 8
Fricks, W., 93

Galaxies, 2
Galileo, 5, 7
Gauges, electronic, 53
GC, 95
General Catalogue of Double Stars within 120° of the North Pole (Burnham), 10
General Catalogue of 33,342 Stars for the Epoch 1950 (Boss), 95
General Catalogue of Trigonometrical Parallaxes (Jenkins), 106
Giacobini, 16
Giant stars, xv, 158, 191–194
Giclas, H. L., 106
Gliese, W., 106, 158
Gravitation, 1, 178–179
Greeley, Frances M., 20
Groombridge, 94

Halley, Edmund, 94
Hamburg-Bergedorf Observatory, 95, 96

Harris, D. L., 158, 167
Harvard College Observatory, 107, 157
HD, 107
Heintz, Wulff, D., 6, 16
Henry Draper Catalogue, 107
Herculis, ζ, 21, 179
Herschel, John, 7–8, 14
Herschel, William, 5, 7, 8, 46
Hertzsprung, Ejnar, 65
Hough, G. W., 9
Hussey, William J., 10–11, 95
Huygens, Christian, 5

Identification of stars, vii, xv, 92–107
Image intensity, 40–45
Index Catalogue of Visual Double Stars, 20, 190
Innes, R. T. A., 14
Interferometers, xiv, 56–57, 59, 61, 67–70, 85
 automatic, 67
 intensity, 70
 speckle, 67–70
International Astronomical Union, ix
Isophotes, 35–39

Jamin compensator, 57
Jeffers, Hamilton M., 20, 57
Jenkins, L., 106
Johannesberg Observatory, 14
Jonckheere, Robert, ix, xiv, 11, 12, 19, 49, 202
Journal des Observateur, 100
Jupiter, 78

Kepler's third law, viii, 148, 158, 179
Kitt Peak Observatory, 16, 69
Kopff, A., 93
Kuiper, Gerard P., 12, 15
Kui 23, 176

Labeyrie, A., 67–70
Laboratoire d'Astronomie, University of Lille, 202
Lacaille, Nicolas, 94
Lagarde, I., 100
Lalande, J. de, 94, 95
Lamont-Hussey Observatory. *See* Bloemfontein Observatory
La Plata Observatory, 11
Laques, P., 66
Lembang Observatory, 19
Leonis, ω, 32
Lick Observatory, 6, 10–13, 20
Lippincott, S. L., 155
Lowell Observatory, 106
Lowell, Percival, 9
Luminosity, 6, 20–21
 and class of star, 191–199
 and mass, 158–166, 167–169, 181
Lunettes et télescopes (Danjon and Couder), 28
Luyten, Willem J., 15, 22, 106
Lynds, 69
Lyot-Camichel double-image micrometer, 63

McAlister, H. A., 69
McDonald Observatory, 16
Magnitude, 30, 33
 absolute, 116, 181, 191, 192, 194–196
 bolometric, 167–169
 catalogues of, 92–93, 158
 and classification, 12, 21, 156–166, 190–196
 and mass, 158–169
 measurement of, 63, 68–69, 74
Mars, 9, 44, 78
Micrometers
 comparison-star, 19, 56
 double-image, 19, 37, 59–63, 85
 filar, xiv, 8, 37, 48–56, 63
 half-wave, 19, 57–59
 interference, 56–57
 screws of, 51–53, 62
 threads of, 49–50, 85, 88, 98
Millburn, 12
Mizar, 5
Montanri, 5
Moon, 178
Moore, C. E., 166
Moreaux, Abbé, xiii
Morel, P. J., 64, 140, 152
Mount Palomar Observatory, 16, 21, 22
Muller, Paul, xiv, xv, 13, 19, 22, 59–63, 135, 141, 189–190
Munich Observatory, 19

Narrabri Observatory, 70
Near-sighted observers, 29–30
Nebulae, 1, 2, 42
Neutron stars, 158
New General Catalogue of Double Stars within 120° of the North Pole (Aitken), 11
NFK, 93
Nice Observatory, ix–x, xii, 6, 16, 87, 202
 Center for the Reduction of Astronomical Plates (CDCA), 37
Number of binaries, 4–5, 174–175

Observation, techniques of, xiv, 85–90
Occultation xiv, 70–71, 96
Occultation Newsletter, 70
Opposite points, method of, 133–145
Orbits, 5, 7, 22, 167
 apparent, 113, 124
 computation of, vii–viii, 110–148, 179–180
 distribution of, 175, 184–188
 of dwarf stars, 12, 13, 182, 184
 eccentricity of, viii, 111, 124, 128, 141, 174
 and mass, 167–168

nodes of, viii, 113–115, 128
period of, viii, 12, 15, 114, 122, 179–180, 194–199
plane of, viii, 4, 113, 116, 125, 174
and range of instruments, 180–187
rapidity of, 196
and semiaxes, 111, 114, 125, 140–141, 144, 179, 181–184, 195
true, 110–113, 116, 124
Orion, trapezium in, 5
Orionis, η, 69

Palace of Discovery (Paris), 89
Parallax, x, 7, 147, 179
 catalogue of, 106, 158
 computation of, xv, 119, 154, 155, 167–168
 dynamical, 166–171, 174
 and mass, 153, 167–171, 174
 and proper motion, 191–194
Paris Observatory, 16–19, 97, 99–100, 202
Pecker, Jean-Claude, xiv
Pegasi
 13, 190
 85, 21
Periastron, 110–115, 120, 124, 128, 135, 141, 196
Perrotin, 16
Persei
 γ, 69
 τ, 68
 12, 69
Peters, J., 93
Photocenter, 35, 37–46, 64, 152
Photoelectric scanning, 71–74
Photography, 22–33, 63, 188
 and catalogues, 100
 and discovery of double stars, 14–15
 with electronic cameras, 66
 and image intensity, 44
 with optical cameras, 64–66
 and stellar mass, 152–166

Photometers, 3, 28, 70
Piazzi, Giuseppe, 94
Pic-du-Midi Observatory, 63, 66
Pickering, E. C., 157
Planets, 44, 65, 71, 78, 94, 155
Pleiades, 71, 178
Pogson's formula, 166, 168, 181
Position angle, 46, 47, 50, 72, 88, 121–122
Positiones mediae (Struve), 8
Precession, 98
Prisms, 53, 56, 59–60, 67, 122
Procyon, 21
Proper motion, x, 15, 21, 96, 154, 155, 191–194
 catalogues of, 20, 106
Pulkovo Observatory, 9
Pulsars, 158

Rabe, 19
Radial velocity, 119, 125
 and stellar mass, 149–166
Rakos, Karl D., 71
Red dwarfs. *See* Dwarf stars
Refraction, 81, 82
Residuals, 133, 145–148
Riccioli, 5
Rigel, 174
Right ascension, 92, 98–100, 106
Romani, L., 166, 169
Rossiter, R. A., 14, 22
Ross 614, 15
Russell, H. N., 166

SAO Star Catalogue, 96, 107
Saturn, 44, 78
Schatzman, Evry, viii
Schiaparelli, Giovanni, 9
Schlesinger, Frank, 95
Schoenfeld, 97
Schuster's effect, viii–ix
Scorpii
 δ, 69
 ξ, 40

256 Index

Secchi, Father, 157
Secondary spectrum, 32, 86
Seeing. *See* Atmospheric distortion
Semiaxes
  and mass, 148–149, 153, 158
  and orbits, 111, 114, 125, 140–141, 144, 179, 181–184, 195
  and separation, 195
Separation, 3, 4, 21–22, 186, 190, 191
  and image structure, 34–40
  measurement of, 46, 140–142, 179, 195
  and occultation, 70–71
  and photography, 64–65
Sextuple stars, 5, 177
Shadow bands, 83
Sidereal time, 85
Sirius, 3, 6, 8, 21, 156, 157, 175
Sky background, 42–44
Smithsonian Astrophysical Observatory, 96
South, James, 7–8
Southern Catalogue Looseleaf Mimeograph (SDS), 14
Southern Hemisphere, 7–8, 14, 22
  catalogues of, 19, 20
Spectra, 65–66, 169–170, 184
Spectroscopy, vii–viii, ix, 28, 68–69
  and catalogues, 107, 157
  and separation, 3, 4
  and stellar masses, 149, 169
Sproul Observatory, 15, 16, 155
Stars. *See also names of stars*
  clusters of, 178
  colors of, 156–158, 175
  diameters of, 1, 71
  distance between, 2, 178. *See also* Separation
  dwarf. *See* Dwarf stars
  giant. *See* Giant stars
  mass of. *See* Stellar mass
  pressure of, 158
  temperature of, 1, 156–158, 167
  types of, 20, 21, 96, 157–158, 167, 182, 191–192
  volume of, 1, 156, 158, 175
Stellar mass
  computation of, vii, viii, xv, 1, 12, 37, 110, 148–171
  and luminosity, 158–169, 181
  and photography, 164–165
Stellarum duplicium et multiplicium measurae micrometricae (Struve), 8
Strand, Kaj Aa., 65, 106, 155, 158
Struve, Otto, 9
Struve, Wilhelm, 8–9, 48
Sun, viii, 158, 175
Surveys, 9–14
Sydney Observatory, 14
Symms, 56

Telescopes
  care of, 79–81
  coupled, 69–70
  eyepieces of, 29–30, 44, 51, 53, 79, 88
  filters of, 65, 82
  focal lengths of, 32, 50, 51, 86
  focal planes of, 29, 30, 34, 49, 72
  focal ratios of, 30, 32
  and image intensity, 40–45
  magnifications of, 30, 33, 42, 88
  mounting of, xiv, xvi, 7, 8, 46, 79–81
  ranges of, 180–187
  reflecting, 7, 28, 32
  refracting, xiii–xiv, 5, 6, 8, 9–10, 28, 78, 86
  resolving power of, 28, 33, 34, 44, 58
Terrestrial observation, 42–44
Thiele-Innes elements, 116–120, 134, 140
Traité d'astronomie stellaire (André), 95

Ursae Majoris
 α, 3
 ξ, 177
 W, 3
U.S. Naval Observatory, 106, 155
U.S.S.R. Academy of Science, 116

van Biesbroeck, Georges, ix, 16, 19
van de Kamp, Peter, 64, 152, 155
van den Bos, W. H., ix, 14, 19, 20, 125–133, 149
Vega, 156
Venus, 78
Virginis
 α, 70
 γ, 3, 7, 149, 199
 η, 69
 θ, 69
Voûte, J., 19
Vyssotsky, A. N., 12–13

Washburn Observatory, 10
Washington Observatory, 9–10, 16, 87
 card index of, 20, 22
White dwarfs. See Dwarf stars
Wickes, W. C., 67
Wind, 77–79
Worley, Charles E., 12–13, 16, 20, 158

Yale Catalogue, 95, 96, 106
Yerkes Observatory, 6, 9, 10
Young's slits, 57, 67